THE FACTS OF LIFE

The Facts of Life

Shattering the Myth of Darwinism

RICHARD MILTON

FOURTH ESTATE · London

First published in Great Britain in 1992 by
FOURTH ESTATE LIMITED
289 Westbourne Grove
London W11 2QA

A catalogue record for this book is available from the British
Library.

ISBN 1–85702–027–8

Printed in Great Britain by Cambridge University Press

The great tragedy of Science –
the slaying of a beautiful hypothesis by an ugly fact.

THOMAS HUXLEY

Contents

PART FOUR: CREATION

AFTERWORD: *An Evolutionists' Apocrypha*

List of Illustrations

rocks are known to be modern – only 190 years old.

6. An astronaut's footprint in the moondust. Fourteen million tons of meteoric dust enters the Earth's atmosphere each year. If the earth were 4,600 million years old, there would be a layer of dust 180 feet thick over the surface, and a comparable amount on the Moon. In fact there is merely an inch or two of dust on the Moon as this photograph shows.

7. An *Archaeopteryx* fossil. *Archaeopteryx* is said by Darwinists to have evolved from dinosaurs called 'coelosaurs' and to be ancestral birds. But coelosaurs did not have collar bones while *Archaeopteryx* does. And while birds' wings are composed of the 2nd, 3rd and 4th fingers of the hand, *Archaeopteryx*'s wing is composed of the 1st, 2nd and 3rd fingers.

8. One of the best-known 'evolutionary sequences', that of early horses. Although the diagram shows an unbroken line, there are major gaps in the sequence, for example between *Eohippus* and its supposed ancestor, and between *Eohippus* and its supposed decendant, *Miohippus*.

9. Fossil ammonities from the lias of Blockley: *Liparoceras* and *Androgynoceras*. Darwinists have variously claimed that *Liparoceras* is the ancestor of *Androgynoceras*; that *Androgynoceras* is the ancestor of *Liparoceras*; and that the two forms are male and female of the same species. Darwinist theory can accommodate all three conclusions.

10. Fossil ammonites from the gault clay of Kent: *Douvilleiceras*, *Beudanticeras*, *Hoplites*, *Euhoplites*, *Anahoplites*, *Dimorphoplites* and *Mortoniceras*. Darwinists believe they are an evolutionary sequence but there are no intermediate species in the beds between.

11. Marsupial flying phalanger and placental flying squirrel; marsupial Jerboa and placental jerboa; skulls of marsupial Tasmanian wolf and placental wolf. The Australasian marsupials are among many species identical with their placental counterparts. Darwinists believe these species have evolved in parallel but quite separately over the last 65

million years from a common shrew-like ancestor. By random mutation?

12. Restoration of 'Piltdown Man'; restoration of *Pithecanthropus erectus*; restoration of Neanderthal man. Darwinist restorations based on fragmentary finds of bones and teeth always manage to convey a distinctly 'missing link' quality to their former owners. The convincing Piltdown Man is wrongly based on a simple forgery associating a normal human skull with the jaw of an ape. But the same 'artistic skill' and imagination have been applied to the genuine fossils.

Introduction

By the river Thames at Teddington, west of London, straggles a cluster of nondescript factory buildings that seem an unlikely home for a national treasure. Yet the buildings are of national importance to everyone in the United Kingdom because of a few pieces of metal kept there.

The buildings are those of the National Physical Laboratory, and the pieces of metal are platinum standards kept at strictly regulated temperature in air-conditioned chambers, to act as the indisputable authority for our national weights and measures: the pound and the kilogram, the yard and the metre.

Paradoxically, these standards are never used for everyday purposes. No one goes to Teddington to measure out a metre of cloth or a foot of parcel string. But their mere existence – unchanging and unchallengeable – is the nation's guarantee that standards exist: that should it ever be necessary, it is theoretically possible to make a physical comparison with the accepted measure and say with absolute certainty that the subject under test either is or is not of the stated weight or length.

It's not easy for members of the public to visit the National Physical Laboratory, because it is a busy government research establishment and the repository of many secrets. But it is certainly possible. Those fortunate enough to be given access to the NPL will see the famous platinum kilogram standard and the atomic clock that is the authority for Greenwich Mean Time. You can check your wristwatch and, in principle at least, can check up on your 12–inch ruler and the weights of your kitchen scales.

A few miles to the east of Teddington stands the very much more imposing structure of the British Museum of Natural History, famous to generations of school children for its dinosaurs and dramatic reconstructions of the Earth's geological history. This building too is the repository of a 'national standard' but one that is not on display in a glass case and that has proved very much more difficult to track down. The museum is the primary source or authority for the general theory of evolution by natural selection, the theory that is taught in schools and universities the world over. Like millions of people in Britain, I have visited the museum many times to stare in wonder at its contents. But I have been unable to see with my own eyes the decisive evidence for the synthetic theory of evolution. I have been able to see many marvels and to study mountains of evidence: the Geological Column that reconstructs the geological and biological history of the Earth; the dinosaur skeletons and myriad other fossils; marvels like the skeleton of *Archaeopteryx*, seemingly half bird, half reptile; the reconstructed evolution of the horse family. But unlike its counterpart at Teddington, the museum is unable to exhibit the unchallengeable authority that conclusively demonstrates that evolution by natural selection has taken place and is established as fact.

This is very far from saying that scientists have failed to make the case for neo-Darwinist evolution. On the contrary, no rational person can visit the Natural History Museum and not be deeply impressed by the evidence that has been painstakingly assembled. Evidence of historical development over geological time; of similarity of anatomical structure in many different species; of change and adaptation to changing environments. But, frustratingly, even with all this evidence, it is impossible for the genuinely objective person to say, 'Here is the conclusive scientific proof that I have been looking for.'

My disappointment among the glass cabinets of Kensington was the beginning of a ten-year journey to try to find that conclusive proof. It has been an Odyssey that has taken me far both geographically and in time. As a science reporter I am used to beating a path from library to museum to laboratory and back again. Now I had to become a scientific historian as well and even a scientific detective to find the evidence I needed to examine afresh. That evidence is of many different kinds and is found in incongruous and often strange settings: in remote quarries and coastal cliffs; in libraries and museums and even in bank vaults. Some can be seen only with the aid of the electron microscope and some has proved not to exist at all.

But although this book has taken me a decade or more to research, my immediate reason for writing it is a simple personal dilemma. My daughter Julia has just celebrated her ninth birthday. She is fast developing an interest in natural history and like many nine-year-olds is an avid fossil collector and dinosaur spotter. She is beginning to have serious science lessons at her school and is just now being introduced to the idea that life on Earth has arisen spontaneously from a common ancestor in the remote past, and that all the species of animal and plant alive today have evolved from earlier species. Over the next ten years, she will be taught that the mechanism governing this process is that of genetic mutation and natural selection – the neo-Darwinist or synthetic theory of evolution.

The fact that Julia is beginning to absorb the general theory of evolution has been giving me sleepless nights. Is Julia being taught the truth? Or is she – and are we – being seriously misled?

Let me make it clear from the outset that my doubts about the general theory of evolution do not arise from religious objections. I want my daughter to have access to the fruits of scientific enquiry, whatever those findings should prove to be. But I am seriously concerned, on purely rational grounds, that generations of school and university teachers have been led to accept speculation as scientific theory and faulty data as scientific fact; that this process has accumulated a mountainous catalogue of mingled fact and fiction that can no longer be contained by the sparsely elegant theory; and that it is high time the theory was taken out of its ornate Victorian glass cabinet and examined with a fresh and sceptical eye.

My doubt about the theory arises from a number of sources. It comes first and most importantly from the inability of Darwinists to pass the simple test described earlier: to show a thinking member of the public conclusive scientific evidence to substantiate the theory, in the same way that the National Physical Laboratory can demonstrate physical constants, the College of Surgeons can demonstrate the circulation of the blood, or the Greenwich Observatory can demonstrate the expanding universe.

Second, it comes from the world of scientific investigation itself, a world that I regularly write about in my job as a science reporter and where many discoveries have been made in the last two decades that have an important bearing on evolution theory, but have received little publicity.

Third, and perhaps most eloquently, it comes from the disarmingly direct questions of a nine-year-old child: Where did we come from? How old is the Earth? Do the butterfly and the elephant really have a common ancestor?

Today, the neo-Darwinist or synthetic theory of evolution enjoys unrivalled prominence as the only rational theory to account for the origin of species and the evolution of all creatures including mankind. It is the only theory taught in secondary schools and in universities and colleges. It is presented as fact in museum displays, lectures and publications. A few controversial points are referred to in British Museum publications and biology textbooks, but these are viewed as peripheral controversies, whose outcome cannot alter the basic truth revealed by neo-Darwinism. The synthetic theory is universally taken as having been confirmed in all its main essentials – only a few isolated details remaining to be tidied up by specialists in esoteric disciplines such as microbiology.

The teachings of this theory are familiar to everyone educated in a western country in the past fifty years. The Earth is of immense antiquity, formed 4,600 million years ago: life on this planet is also of very great age; emerging spontaneously in ancient seas 3,000 million years ago; and the great variety of species that exist today are all descended from one or a few primitive organisms evolved in those ancient oceans, by a process of random genetic mutation coupled with natural selection. These ideas are the cornerstones of modern historical geology and of our contemporary world view, as familiar to the elementary school pupil as to the postgraduate student of biology or geology.

To most students, teachers and even some scientists it will come as a surprise to learn that recent research into the age of the Earth has produced evidence that our planet may be much younger than had previously been thought – evidence in one case pointing to an age as young as 175,000 years; that existing methods of geochronometry such as uranium-lead decay and radiocarbon assay have been found to be deeply flawed and unreliable; that the extent of genetic change by selection has been found experimentally to be finite; that only a catastrophist model of development can account for important Earth structures and processes such as continental drift and most fossil-bearing rock formations – most of the Earth's surface in fact. These major discoveries have had profound consequences for the neo-Darwinist theory of evolution, yet few of them have found their way into the public domain, still less into school or university textbooks or museum displays.

This book attempts to make accessible, and put into context, these new discoveries to enable non-scientists to evaluate for themselves the status of the general theory of evolution, and the new light cast on existing theories by the latest discoveries.

A number of books attacking the neo-Darwinist theory – and evolution in general – have been published in recent decades by religious writers seeking to promulgate biblical creation as an alternative viewpoint, and I should make it clear at the outset that this book is not in any sense a contribution to creationist literature, nor does it represent the creationist viewpoint (although some creationist objections to neo-Darwinism that have a basis in scientific research are included here).

Because some of the scientific matters discussed in this book are highly controversial, I have included references to original studies wherever necessary. However, responsibility for the conclusions drawn from these sources is mine alone.

PART ONE

Chaos

CHAPTER 1

Through the Looking-Glass

When a fair-skinned woman lies on a sunny beach for any length of time, her skin will acquire a tan in response to the Sun's ultra-violet rays. The longer she lies on the beach, the darker her tan will become. But no matter how long she lies on the beach, her children will never be born with her suntan.

This everyday experience is at the very heart of the approach to evolution taken by Darwin and his successors. The search to understand such apparently simple observations has led to a theory of great sophistication whose power of explanation is formidable. But it is a theory that runs counter to intuition, that has proved impossible to confirm experimentally, and that Darwinists and their opponents have fought tooth and nail over for more than 150 years.

Darwinists believe that the reason it is impossible for a child to inherit characteristics its parents acquire during their lifetime (such as the mother's suntan) is that evolution is not supervised by any directing force or design, but by chance operating together with natural selection. In the example given, this means that the black and brown people of the world have not acquired dark pigmentation because they live in sunny regions or because it would be useful to them to have dark skin to screen the Sun's rays. Instead, chance alone has given one or a few ancestors darker skin and natural selection favoured the survival of the darker-skinned people because they lived in sunny regions.

Possibly a darker-skinned person was born in northern latitudes from time to time, but his or her skin gave no special survival advantage and any offspring did not prosper especially so the predominant skin tone remains white.

Modern Darwinists have actually taken this idea one step further and added a codicil to the theory from the field of microbiology. The reason that acquired characteristics cannot be inherited, many believe, is because the mechanism of inheritance – the genes contained in our sexual cells – cannot be constructively affected by the environment. The genetic code is a one-way system. Information can be read out when a new life is generated but information cannot be written in to alter the characteristics of that new life.[1] If the offspring differs at all radically from its forebears, believe Darwinists, it is because of random chance and nothing more. The fundamental mechanism of evolution is, in Professor Jacques Monod's memorable phrase, 'chance and necessity.'[2] In the Darwinian world it is possible to look into the mirror of genetics and even to read what is written there, but, unlike Alice, we can never go through the looking-glass.

The Darwinian idea is powerful and beautiful. It has sustained evolution theory for a century and a half. Many discoveries after Darwin's death have tended to confirm the idea. And many new ideas have been advanced concerning, for instance, the origin of life from non-living materials that make use of the Darwinian concept and that stand head and shoulders above any competing theory.

And yet many people, both scientists and laymen, have entertained nagging doubts. Do we really believe that black people are black by accident? What kind of accident was it? Why don't we see such accidents happening today? Why does the fossil record not show us such accidents happening in the past? Once the questions begin, it is difficult to know where to stop.

For instance, if we don't see genetic mutations – the accidents of inheritance – because they are very rare, then how can there have been enough of them to produce anything as complex as humans? Darwinists say this is because many billions of years have elapsed since the Earth cooled. The geological strata covering the Earth's surface have taken millions of years to lay down and the fossil creatures in them lived millions of years ago. But if these rocks take millions of years to form, why do we find trees, forty feet tall, in the vertical position of growth in coal seams?[3] And why are there still comets circling the Sun that should have dissipated in only tens of thousands of years after the Earth's formation?[4] And if nature can produce such rich diversity as the present animal and plant kingdoms by pure chance, why is it that thousands of years of serious guided selection by mankind has resulted only in trivial sub-specific variation of domestic plants and animals, while not one new species has been created?

These and hundreds of similar questions begin to press for answers. But before the Darwinian response to those questions can be evaluated, it is necessary first to gain a firm grasp of just exactly what modern Darwinists do believe, and perhaps more importantly, how they came to believe it.

The story begins appropriately enough in a garden – not that of Eden but rather the Botanical Gardens in Ghent where, in 1898, a 27-year-old Austrian botanist called Erich Tschermak began to take an interest in breeding garden peas. After only two years work, trying to breed distinctive characteristics, he found to his astonishment that the hybrids showed a mathematically precise ratio of yellow-seeded to green-seeded peas. Reading the literature on peas he found a cross reference that seemed interesting and in 1900 he sent to the library of Vienna University for the papers.

At Amsterdam University, the Professor of Botany, Hugo de Vries, made a similarly exciting discovery in 1886. He found certain wild varieties of the evening primrose that differed markedly from the cultivated variety. Coining the term 'mutation' to describe the phenomenon, de Vries started a long series of plant-breeding experiments to see if he could breed mutations. In 1900, he made a breakthrough that started him searching through the university's literature on pea breeding.

In 1892, the instructor of botany at the University of Tübingen, Germany, Carl Erich Correns, conducted some research that required him to breed garden peas, and in 1900 he found the same mathematical pattern that his two contemporaries had discerned in their experiments. He too searched the literature in the university library, and like his colleagues was astonished to discover that the entire subject had been researched and published in great detail a generation before by an unknown Augustinian monk, Gregor Mendel.

The simultaneous independent discovery by the three men – all of whom later had distinguished careers in biology – came at a time when Darwin's theory had been all but consigned to the scrapheap of history. The theory had tottered on for a few years following Darwin's death but had fallen out of favour because it lacked a credible mechanism that could cause change to take place in the species that populate the world. Darwin had suggested microscopic fluctuations in form, gently edging a species in one direction rather than another: the giraffe's neck getting imperceptibly longer with each generation. But this phenomenon was nowhere observed in nature. Stability is the norm, not change – however slow – and Darwin's idea of 'heritable characters' was simply a non-starter.

When Hugo de Vries found his markedly changed evening primroses in the wild, he immediately surmised that he had found such a mechanism: not the trivial 'fluctuations' of Darwin but substantial 'mutations' that accounted for bigger, sudden changes in form. These mutations must be caused by radical changes in the basic blueprint of heredity – what were later called genes. What de Vries and his co-discoverers had observed in pea plants was the tendency of certain genetic characteristics to dominate in a majority of offspring – for most of them to be tall when a short pea was crossed with a tall one. And it was this property, today called the first law of genetics, that Gregor Mendel had discovered with his pioneering efforts in the monastery garden in the 1860s.

It has often been remarked as strange that Mendel's great achievement was never realised or acknowledged in his lifetime. His paper was circulated to all Europe's great libraries and was read by many eminent biologists, none of whom saw the importance of his discoveries. But if the European mind was unreceptive in the 1860s, it had become highly receptive four decades later, and Darwinian evolution theory was rescued from the scrapheap at a single stroke. Combined with Mendelian genetics, and the concept of mutation, Darwinism re-emerged with a solid experimental foundation as the neo-Darwinist or synthetic theory of evolution.

In the neo-Darwinist theory, species evolve into other forms by means of natural selection as Darwin had suggested. But they do so not because of the trivial variation that occurs between all individuals but because of random mutations in their genetic makeup, most of which are lethal, but a few of which favour survival. Thus blind chance combines with necessity to shape the animal and plant kingdoms.

From the rediscovery of Mendel until today, the synthetic theory has remained pre-eminent as the only scientific theory to account for the origin of species and the evolution of all creatures including mankind. No other theory is taught in secondary schools (with the possible exception of the few religious schools that remain outside the state system) or in universities and colleges.

In the United Kingdom, for instance, the new National Curriculum for schools lays down these instructions for teachers of biology: 'Pupils should develop their knowledge and understanding of variation and its genetic and environmental causes and the basic mechanisms of inheritance, selection and evolution.' The National Curriculum's Attainment Target on 'Genetics and Evolution' specifies pupils' objective as 'Understanding the relationship between variation, natural selection and reproductive success in organisms and the significance of their relationship for evolution.'

So strong has the Darwinian model of evolution become that it has vanquished and displaced once and for all the lingering challengers it once had: Lamarckism or the inheritance of acquired characteristics; various versions of vitalism – the idea that evolution is supervised by some non-physical natural force – and of course the biblical account of creation by an almighty hand. It is unsurprising that the principal opponents of evolution theory, both in its original form as conceived by Darwin and in its modern synthetic form, have been individuals who accept the biblical account of creation, who see Darwinism and neo-Darwinism representing an assault on their religious beliefs and who wish to have the theory given much reduced prominence, especially in education.

The movement has been active for most of this century, especially in the United States. From the Scopes trial in Tennessee in 1925 to Creationist pressure groups of the 1990s, religious fundamentalists in America have repeatedly challenged evolutionists whatever their hue, sometimes successfully.[5] Some Creationists are also scientists – a few of considerable standing academically – and they have offered substantive scientific criticisms of synthetic evolution (some of which are included in this book).

An important factor in bringing about the universal dominance and acceptance of synthetic evolution has been that virtually every eminent professional scientist appointed to posts in the life sciences in the last 40 or 50 years, in the English-speaking world, has been a convinced synthetic evolutionist. These eminent names include men such as Sir Gavin de Beer (Professor of Embryology at University College London 1945–50 and Director of the British Museum of Natural History 1950–60), Sir Julian Huxley (Professor of Zoology at King's College, London University, and Secretary to the London Zoological Society), J. B. S. Haldane (Professor of Genetics at London University 1933–37 and Professor of Biometry at University College 1937–57), and C. H. Waddington (Professor of Biology at Edinburgh University). In the United States, leading synthetic evolutionists have included Ernst Mayr (Professor of Zoology at Harvard University 1953–61 and Director of the Museum of Comparative Zoology 1961–70), Theodosius Dobzhansky (Professor of Zoology at Columbia University 1942–60), and George Simpson (Professor of Palaeontology at Columbia University, Professor of Palaeontology at the Harvard Museum of Comparative Zoology 1958–68, Professor of Geosciences at the University of Arizona).

One of the most spirited of Darwin's champions from the non-English-speaking world has been France's Nobel prize-winning microbiologist and Director of the Pasteur Institute, Jacques Monod, whose 1970 book *Chance and Necessity* caused something of a shock wave on both sides of the Atlantic for its uncompromising portayal of life as no more than chemistry and statistics.

These men, as well as occupying powerful and leading academic teaching positions were also prolific and important writers whose influence has been widespread in forming the consensus. In Britain, Darwin's theory has been almost a family business for the Huxleys: Thomas Huxley acting as Darwin's champion, and grandson Sir Julian becoming an equally eminent biologist. Sir Julian's 1942 book *Evolution – the Modern Synthesis*, revised in 1963, is probably the closest published work to a complete all-embracing textbook on synthetic evolution (and is an essential starting point for any enquirer on the subject).

In setting out to criticise neo-Darwinism, I have made these writers my chief sources of authority for articulating the modern synthetic theory, in order not to misrepresent the current neo-Darwinist position. I have quoted primarily from their works (as well as from the writings of some of their close colleagues and adherents) so that the beliefs and ideas described are accurately representative rather than merely easy targets for refutation.

To dissent from the dominant scientific idea of the life sciences in the twentieth century may seem both foolhardy and unnecessary. After all, how could so many important scientists be wrong? Surely only religious cranks question evolution – the earnest kind who want to sell you strange newspapers and eagerly seek conversations 'about life and its meaning'?

My reason for setting out to re-evaluate the received wisdom of synthetic evolution is that something pretty nearly the opposite of these sentiments is actually the case. Far from being the province of cranks, it is the non-Darwinian view that is supported by modern findings. A very few (albeit very able and distinguished) scientists were responsible for making the synthetic theory pre-eminent in the natural sciences. They were able to do this in an era when intellectual authority often counted as much as experimental accuracy or innovation. And the principal findings that undermine the synthetic theory have come from a new generation of scientists, less concerned with authoritarian theories and more concerned with unravelling mysteries.

Their findings have arisen from research in every one of the complex interlocking set of disciplines that go to make up the synthetic theory: geology, stratigraphy, petrology, radioactive dating, palaeontology, comparative anatomy, zoology, genetics, microbiology, organic chemistry. These findings undermine and challenge very many fundamental tenets on which the theory is constructed: tenets as elementary as the age of the Earth; the formation of sedimentary rocks and the formation of the main features of the Earth's crust; the limits to specific variation; the causes of extinctions; and even the possible origins of life – long considered settled in broad outline. Yet these new findings have been given short shrift by the ruling ideology of science, primarily because of the huge investment of time, money, resources and scientific reputations that neo-Darwinism represents.

But of course, there is much more to science's commitment to neo-Darwinism than careerism. It is an elegant, comprehensive, rationally based and economical set of neatly interlocking ideas that provide an important basis for understanding one of the most mysterious areas of scientific study – the origin and development of life.

I said earlier that I intend taking the theory out of the glass cabinet in which it is so reverently kept and looking at it a little less reverently and a little more closely. I want to begin this examination with a closer look at what is probably the central issue: the age of the Earth. The reason that this issue assumes key importance is because the central mechanism of neo-Darwinism, genetic mutation, means that change has to take place at an agonisingly slow pace – requiring hundreds of millions or even billions of years. If the Earth is of such an age, then neo-Darwinism could be true. If the Earth is not of such an age, then the theory cannot be true – despite what other evidence may indicate.

CHAPTER 2

A Matter of Conjecture

So, how old is the Earth? On the face of it, the answer to this question is cut and dried and no longer relevant to a discussion of evolution. It seems irrelevant to the modern debate because the age now universally accepted for the Earth is so vast – 4,600 million years – as to allow life to have evolved not once but many times. But let us use our imaginations for a moment to ask two heretical questions. Does an age of 4,600 million years really provide enough time for evolution to have worked along Darwinian lines? And – even more outrageous – what if the Earth is not so old as we think?

Try this thought experiment on the first question. What has to happen in order for life to get started in the primeval oceans, and to develop by mutation and natural selection into the animal and plant kingdoms we see today? First, the inert chemicals in the sea have to form amino acids, probably under the influence of ultra-violet light and electrical discharges in the form of lightning. This process was demonstrated by Harold Urey and Stanley Miller at Chicago University in 1953.[1] In step two, the first amino acids in the early ocean have to combine to form the stuff of life, protein molecules. It is these giant and complex molecules that ultimately constitute plant and animal life, and although the mechanism by which they might have formed spontaneously is not known and has not been demonstrated in the laboratory, it is not impossible to grant that it could have occurred given enough time. The third step will be the explosive variation and growth of all manner of life forms based on protein, from bacteria to Beethoven. Given steps one and two, this is not impossible to imagine, and from a Darwinian standpoint it would perhaps be surprising if it did not happen.

It is steps two and three that have a bearing on the age of the Earth. Although step two, the spontaneous formation of protein molecules, is an unknown process, it is theoretically possible to assess how long it would take to happen by chance. Based on the size and complexity of such molecules, Murray Eden, Professor of Electrical Engineering at MIT, calculated that a very simple synthesis would be expected to happen by chance once in about 1,000 million years.[2] On the face of it, even these very lengthy odds can easily be accommodated in the 4,600 million years that most geologists assign to the Earth's history. But look a little closer.

Life is not spontaneously emerging today in the seas. This is attributed by evolutionists to the fact that conditions have changed since life evolved in the archaic oceans.[3] So how long exactly were the conditions suitable for this spontaneous happening? The time available would be bounded by two events. The cooling of the Earth and the establishing of the oceans would be the earlier marker event. This is said to have occurred about 3,800 million years ago (the date when the first surface water appeared).[4] The upper marker would be the date of the first fossil of a living thing. Just where this upper marker should be placed is a controversial matter. The conservative view is that the first sign of life is represented by organisms called *Eobacterium isolatum* and *Archaeospheroides barbertonensis*, that are dated from about 3,200 million years ago.[5] This gives us a window of opportunity for the spontaneous occurrence of the first microorganism of roughly 600 million years. Actually, the gap is smaller than suggested by this crude sum because it would take a considerable time for the new oceans to acquire the right mixture of basic chemicals to make the primeval 'soup', and at the other end, bacteria must have been predated by some simpler non-replicating molecules of which no trace survives. But let us be generous and allow the full 600 million years (what is a few million years when we have so many at our disposal). This interval has to accommodate not only the spontaneous combination of basic materials into amino acids, but also the combination of amino acids into protein molecules, the appearance of at least one self-replicating molecule and the subsequent evolution of this molecule into self-replicating cellular bodies to the bacterial level. And remember that of these four steps, one alone (the second) has been estimated to happen by chance once in 1,000 million years.

So, of the 4,600 million years of geological time that Darwinists have granted themselves, only a small fraction – less than 600 million – is actually available to accommodate the processes they believe to have taken place.

The latest evidence indicates that the gap is even narrower. Geologists Hans Pflug and H. Jaeschke-Boyer studying ancient rocks from Greenland, said to date from 3.8 billion years ago, found fossil cell-like structures in 1979 which they named *Isosphaera*.[6] The fossils are those of a primitive yeast-like organism. In the same year, Australian geologist J. S. R. Dunlop discovered fossil structures formed by microorganisms in Western Australia said to date from as much as 3.5 billion years ago.[7] In 1980, C. Walters, A. Shimoyama and C. Ponnamperuma examined *Isosphaera* for evidence of photosynthetic activity and announced: '. . . we have now what we believe is strong evidence for life on Earth 3,800 million years ago.'[8]

The meaning of these discoveries is clear. If the first surface water formed 3,800 million years ago and the first microorganisms came into existence 3,800 million years ago, then there was zero time available for the spontaneous appearance of life. Life, it seems, did not wait for blind chance to roll the dice, but erupted at the first available instant, leaving Darwinists with no time at all for their probabilistic processes.

Consider now a second searching question. One that is even more heretical. Let us ask what evidence we have for the age of the Earth and what grounds we have for accepting that evidence. The importance of this question, as observed earlier, lies in the fact that an Earth of immense age is indispensably necessary to the neo-Darwinist theory because genetic mutation and natural selection are processes that are conceived of as working very slowly over hundreds of millions of years. If the Earth were only a few million years old then there simply would not have been enough time for natural selection to work. Whether we liked it or not, we would be compelled to seek a fresh explanation for the origin of living species.

On this fundamentally important question, the Natural History Museum and all other modern authorities are in complete agreement. The Earth is 4,600 million years old. What is more, different periods of the Earth's history have been characterised by the formation of different kinds of rock containing the fossil remains of distinctive kinds of creature. These different periods have also been dated to give what is usually referred to as the Geological Column of the Earth's history. The column is reproduced as Figure 1.

By referring to the geological column anyone can tell the age of a rock or fossil that he or she finds. For instance, the white cliffs of Dover consist of chalk dating from the end of the Cretaceous period which, the column tells us, dates from 65 million years ago.

The dates attached to the geological column have been arrived at and refined over the past century or so. The most recent evaluation, and the one quoted in Natural History Museum publications, is that of Van Eysinga published in 1975.[9] This scheme (which is the source of Figure 1) is closely similar to that used in most Western museums and universities since the early decades of this century, and is based on the pioneering work of Arthur Holmes[10] in the UK and Henry Faul in the US.[11] Some minor disagreements may exist among geologists but a very wide measure of agreement exists over the big issue, that the earliest rocks of the column are around 4 billion years old, and over most of the details in Figure 1, for example that the Cretaceous period began around 140 million years ago and ended around 65 million years ago.

When I began to research this question a little more closely I uncovered a puzzle. Those experts I referred to and the authoritative textbooks I consulted all told me that modern dating has been accomplished by using radioactive methods and hence was an absolute dating method of a far higher order of accuracy than all previous methods – most of which relied on calculations involving one or more relative factors. These relative dating methods had relied on such factors as the increasing salinity of the oceans, or the Earth's rate of cooling, and are now considered unreliable.[12] Radioactive dating, though, is used to date the rocks directly and hence was welcomed as an absolute method.

The puzzle arises because radioactive dating techniques can be applied only to volcanic rocks which contain some radioactive mineral – the primary rocks of the Earth's crust. But the geological column consists of sedimentary rocks – rocks formed from sediments laid down on the beds of ancient seas and composed of particles of those primary rocks. So, of course, any age determination made using these particles will be the same as that of the primary rocks from which they were derived. In some common sedimentary rocks, such as chalk or limestone, there are not even particles of the primary rocks present and so radioactive dating cannot be used at all. Happily for Englishmen and women, the white cliffs of Dover are not radioactive.

In *The Age of The Earth* published by the Institute of Geological Sciences, the position is succinctly explained by John Thackray:

> The only sediments which can be dated directly are those in which a radioactive mineral is formed during diagenesis [laying down] of the sediment, such as the rather uncommon illite shales and glauconitic sandstones; other sediments give only the age of the parent rock from which the mineral grains that make them up are derived.[13]

How then did Holmes, Faul and Van Eysinga arrive at the dates attached to the sediments of the geological column? The Institute of Geological Sciences explains:

> Where lavas or volcanic ashes are interbedded with a sediment of known stratigraphic age, then a date may be given to that stratigraphic division. Where an igneous rock intrudes one sedimentary unit and is blanketed by another, then the sediments may be dated from the igneous rock by inference. The rarity of such cases, together with analytical error inherent in age determination, mean that isotopic ages are unlikely to rival or replace fossils as the most important means of . . . correlation.

It turns out that what has been dated by radioactive decay methods is not the sedimentary rocks themselves but the isolated intrusion into them of igneous or primary rocks, usually as volcanic material. This has been a rare and purely fortuitous process and one that is unreliable – so rare and so unreliable that the Institute of Geological Sciences thinks it unlikely to replace or even rival fossils as a method of dating. Nor is this all, for the method depends in turn on a further chain of inference. For the geological column of Van Eysinga is nowhere to be found in nature. It is an imaginary structure that has been synthesised from comparing a stratum of rock in one part of the world with a similar looking stratum in another part of the world (see

Chapter 7 for a more detailed discussion of the composition of the geological column).

Naturalists themselves are often confused in their knowledge of this question. Sir Gavin de Beer, for example, Director of the British Museum of Natural History from 1950–60, wrote in the introduction to the museum's guide to evolution, published in 1970, that the rocks forming the geological column had been dated by radioactive methods:

> Estimates of time based on disintegration of radioactive material enable various levels of evolutionary lineages to be dated and the time measured during which certain changes have occurred, thereby providing quantitative evidence of evolution rates and the duration times of genera and species.[14]

This claim, which is universally believed and taught in schools and universities throughout the world, is entirely wrong. And when Darwinists speak of absolute dating of the geological column and the fossils it contains by radioactive methods they are quite mistaken, there is nothing absolute about it. In fact the method ought to be referred to as 'comparative dating', because it dates the sedimentary rocks by inference alone through their relationship to the rare samples of igneous or primary rocks that *are* being dated.

When I pursued this question a little further, I found that there is in reality another factor that has been used to arrive at the age of the geological column and the fossils it contains – conjecture. This process crept into geological dating at a very early stage when Charles Lyell, the nineteenth century's most prominent geologist and Darwin's mentor in geological matters, attempted to date the end of the Cretaceous period by reference to how long he thought it would have taken the shellfish (whose fossils are found in later beds) to have evolved into their modern descendants. Lyell estimated that the Cretaceous ended 80 million years ago – not too far from today's accepted figure of 65 million, plus or minus 3 million.

According to Harold Levin of Washington University, 'By comparing the amount of evolution exhibited by the marine molluscs in the various series of the Tertiary System with the amount that had occurred since the beginning of the Pleistocene Ice Age, Lyell estimated that 80 million years had elapsed since the beginning of the Cenozoic.'

Levin adds that 'He came astonishingly close to the mark.'[15] In fact, it is not at all astonishing when you know that today's accepted date has been derived not from an absolute, independent source but from conjectures including Lyell's.

The kind of surmise used to supplement the relative dates yielded by radioactive dating includes assumptions about the rates at which sediments are laid down on the bottoms of lakes, seashores and ocean floors; estimates of the rates at which forests are turned into coal deposits; and estimates of the rates at which certain very long-lived families of creatures might have evolved. But though these conjectures are embodied in the modern view of the age of geological deposits, they are rarely if ever disclosed in geological or biological textbooks, and they are rarely exposed to debate.

Curiously, too, no geologist seems to have checked out the geological column dates with an electronic calculator on a common-sense basis. Let us go back to the illustration of the column in Figure 1 and look again at the thickness of the rocks in each period compared with the length of time assigned to those periods. Notice that there is a remarkable consistency between assigned age and thickness of deposit. For instance the Cretaceous period is said to have lasted 65 million years and is 15,000 metres thick – an average annual rate of deposition of 0.2 millimetres. Now look at the Silurian period: this, too, yields an average rate of deposition of about 0.2 millimetres per year – as do the Ordovician, the Devonian, the Carboniferous and the rest. It is only when we come to relatively modern times in the Cenozoic era that rates of deposition vary much, and here they appear to speed up slightly.

This is a very remarkable finding. One naturally expects uniformitarian geology to favour uniformity, but this is too much of a good thing. Throughout widely changing climatic conditions, advancing and retreating oceans, droughts and Ice Ages, the rate of sedimentation appears to remain amazingly constant regardless – throughout the thousands of millions of years that are said to have elapsed. The presumed rate of deposition itself – about the thickness of a human hair in a year – is a matter looked at in more detail later. But it is worth pausing in passing to notice that such a slow rate would be quite incapable of burying and fossilising entire forests, dinosaurs or even a medium-sized tadpole.

The rate of deposition for the geological column is, of course, merely an average. But it is confirmed by other more detailed studies. In 1929, for example, W. H. Bradley conducted a study of varves (layers of sediment assumed to be annual) in the shales of Wyoming's Green River formation. Bradley found that the 790 metres of shale were deposited in 6.5 million years – which again gives about the same average annual sedimentation rate, in this case 0.12 millimetres per year.[16]

Of course, all these sediments, with their time capsule contents of fossilised creatures from the past, were laid down long after the Earth was formed and long after the decisive event took place in the chain of evolution – the origin of life itself in ancient seas. It is the rock from which those later sediments were derived – the primary bedrock of the Earth's crust – in which we are chiefly interested if we wish to date the Earth. The key question remains how old is the Earth? And to examine the answer that has come to be accepted on this score, we must look more closely at radioactive methods of dating.

CHAPTER 3

The Key to the Past?

In the years following the Second World War, American chemist Willard Libby made a discovery which won him the Nobel prize for chemistry, which revolutionised the study of the Earth's prehistory, but which ultimately was to provide unexpectedly disconcerting evidence on the age of the Earth itself.

Libby's discovery was the now famous radiocarbon method of determining the age of organic remains, which gave archaeologists their first practical tool for routinely dating the past. At the time of its discovery and its first application to archaeological sites around the world in 1949, the radiocarbon method appeared to confirm that mankind's past was indeed of great antiquity, and that geologists and evolutionists had been perfectly justified in continually pushing further back in time the dawn of humanity.

Field archaeologists in the 1950s, applying the new power given them by chemistry, confidently assigned absolute dates to early human prehistoric settlements with a precision which must have astounded their teachers of a generation before. The city of Jericho was said to have been a thriving human settlement 11,000 years ago, while neolithic sites in Russia and Africa were dated as being well over 50,000 years old. The author of *Encyclopaedia Britannica*'s article on prehistoric Africa, for instance, says, 'Radiocarbon dating suggests that the Earlier Stone Age may have lingered on until about 55,000 BC.'

The readiness of science today to accept a great antiquity for the Earth and mankind contrasts sharply with the attitude of scientists only a century ago. This radical change in outlook involved the overthrow of the old geological belief in a catastrophic origin for the rocks of the Earth's crust and its replacement by the modern uniformitarian theory – the idea that the rocks have formed slowly over millions or billions of years.

At the time that Darwin set sail for South America in the *Beagle* in 1831, the Earth's age was reckoned merely in thousands of years, and not many thousands at that. One well-known early attempt to date the Earth is that of Archbishop James Ussher of Armagh, a noted Bible scholar who deduced through careful analysis of biblical texts that the Earth was created in 4004 BC. The Archbishop's finding was published in 1650 and soon after was added as a marginal notation to the Book of Genesis in the Authorized Version of the Bible where it remained until Victorian times, and can still be found occasionally today.

A contemporary of the Archbishop, Dr John Lightfoot, Master of St Catharine's College and Vice-Chancellor of Cambridge University, was able to endorse this date and indeed refine it with astounding precision. 'Man was created by the Trinity,' wrote Dr Lightfoot, 'on October 23rd 4004 BC at nine o'clock in the morning.' As Ronald Millar has pointed out, only a Cambridge Vice-Chancellor would have the audacity to assign the date and time of the creation to the beginning of the academic year.[1]

A number of the influential geologists in Darwin's day were also clergymen whose religious views strongly influenced their scientific beliefs. This religious complexion to geology in the eighteenth and early nineteenth centuries – an otherwise flourishing era for rationalist thinking in science – influenced theories of rock formation and the age of the Earth in two important ways.

First, widespread acceptance of the biblical creation story contained in Genesis meant that the cleric-geologists neglected to question how the Earth began or how life originated because they believed they already had the answers to these questions. And second, the creation story constituted a ready-made theory to accommodate all their scientific observations (often meticulously detailed) thereby stifling the formation of any new theory when they discovered new evidence in the field.

When these researchers found thousands of feet of compacted mud-like sediments containing the bones of dead animals, the discovery was taken as clear evidence of Noah's flood described in the Bible. Hence the prevailing geological theory of the pre-Darwinian era was that of catastrophism – the doctrine that the rocks of the Earth's crust were formed more or less simultaneously as a result of a divinely ordained Great Flood.

Some of the attempts by pre-nineteenth-century geologists to fit their observations to biblical teaching appear obviously contrived and rather absurd from our perspective today. Swiss naturalist Johann Scheuchzer, who discovered some early vertebrate remains of a salamander around 1720, exhibited them widely as the remains of the imaginatively named *Homo diluvii testis* – Man, a witness to the flood.[2] (Some believe that Scheuchzer was seeking to turn an honest copper or two with his discovery, in which case we must blame the gullibility of his customers rather than the inadequacy of eighteenth century science.)

In general, though, the observations of nature made at this time were models of scientific accuracy and would do credit to any modern researcher. Unfortunately when the theory of catastrophism fell into disrepute after Darwin, many of the observations of the cleric-geologists were rejected as religiously inspired pre-scientific thinking: observations which did indeed support a catastrophic origin for many rocks. The detailed evidence for catastrophism is examined in a later chapter, but one observation of this type that was well-known in Darwin's day may be mentioned now by way of example: the occurrence of 'graveyards' of millions of land-dwelling (not marine) creatures who suffered death simultaneously.[3]

Darwin and his supporters realised at an early stage that their theory demanded vast reaches of geological time to support the supposed microscopic changes in form from one generation to another. Equally, evolutionists stood in need of a geological basis for this great antiquity – a mechanism that worked slowly and gradually rather than one which worked suddenly and all at once. They rejected catastrophism and instead found the mechanism they sought in an idea taking shape among the new generation of secular geologists who asserted that sedimentary rocks (that is, fossil-bearing rocks) were formed slowly by the same processes that can be seen on the ocean bottom today: the deposition of silt and sand which becomes cemented and compacted over millions of years to form successive strata of rock.

Under the reassuring-sounding label of Uniformitarianism these ideas were actively promoted by secular geologists like James Hutton and later Charles Lyell, who was Darwin's coach on geological issues. The uniformitarian doctrine is summed up in the famous phrase 'the present is the key to the past' – a concept eagerly accepted by Darwinists as ready-made for their theory, and one expounded on at length in Lyell's *Principles of Geology*, the primary geological work of the century, published between 1824 and 1833.[4]

The important point to notice here is that it was the imperative need for great antiquity which deposed catastrophism, rather than any new scientific discoveries or observations – it was a new way of looking at things, not a new piece of knowledge. But superficially, the change in view seemed to be a shift away from naive belief in biblical tales of creation and flood, and towards a newly established scientific viewpoint. And those who continued to argue the case for a catastrophic origin of rocks were seen as merely making a last-ditch attempt to rescue the religious doctrine of the creation as told in Genesis.

Darwinists needed time, and lots of it: uniformitarians had the geological theory which demonstrated great antiquity. Geologists needed a firm foundation for the relative dating and correlation of the many sediments piled one on another in the past – the many strata of the geological column: Darwinists were able to supply the key to the stratigraphical succession of the rocks by comparative anatomy of the fossils contained in those strata, interpreted along evolutionist lines. Thus an unusual academic interdependence sprang up between the two sciences which continues to this day. If a geologist wished to date a rock stratum he asked an evolutionist's opinion on the fossils it contained. If an evolutionist was having difficulty dating a fossil species, he turned to the geologist for help. Fossils were used to date rocks: rocks were used to date fossils.

A modern example of palaeontologists using fossils to date rocks in a circular way is provided by one of the most famous of all North American dinosaur discovery sites, the rocks at Como Bluffs, Wyoming. I only regret that this example involves one of today's most innovative researchers, Robert Bakker of the University of Colorado. It was at Como Bluffs in the 1870s and 1880s that palaeontologists such as Edward Cope and O. C. Marsh discovered more than 120 new species of dinosaur, including diplodocus and stegasaurus. The many strata exposed in the steep cliff at this seminal site have subsequently yielded many more specimens and they are still worked today by scientists from many universities.

Of the site, Robert Bakker says:

> At a place like Como Bluffs you have layer after layer – it's like getting a burst of frames from a motion picture of how the dinosaurs came, flourished and went extinct. At any one place in the world, you don't have the whole history of dinosaurs, in fact you don't have the whole history of one family of dinosaurs, you just have a little burst of fossils.

> We don't yet have radioactive beds that can give us a nice hard number [on the age of the deposit]. But by comparing the fossils we get at the bottom of the section and at the top, it's about 10 million years. So all of this history is played out roughly over about 2 million dinosaur generations, 10 million chronological years.

Ironically, not only is there no radioactive basis for the dating of Como Bluffs, there is, as Robert Bakker says, not even a complete history of a single dinosaur family at the site. Yet we are given the confident assertion concerning the number of dinosaur generations and the number of years to which this sequence is equivalent, with no solid physical basis. No other scientific discipline would be permitted even to consider such procedures, but when palaeontologists date rocks by means of fossils, they do so with the authority of Charles Darwin himself.

This circular process ought to have aroused suspicion, if not among its practitioners, then among scientists of related disciplines. In fact it went unremarked and unchallenged because the discovery and introduction of methods of dating based on radioactive decay in the early years of this century appeared amply to vindicate the Darwinist-uniformitarian view and to justify their interdependence.

In the last two decades, however, further research into these technical methods of dating has revealed a number of worrying inconsistencies in the now orthodox view of the Earth's age: that radioactive dating techniques are far less reliable than was previously thought; that other methods of dating which have been devised tend uniformly to show that the Earth may be much younger than has been supposed by Darwinists; and that nothing like the billions of years required by evolution theory have elapsed since the Earth's formation.

The first clue that something may be amiss with the view of uniformitarian geology and its claim for an old Earth came paradoxically from the technique that seemed most to support that view – Willard Libby's radiocarbon dating method. To appreciate exactly why the radiocarbon technique has had such unexpected consequences, it is necessary first to look at just how the technique was supposed to work.

Radiocarbon – radioactive carbon 14 – is a form of carbon created in the upper atmosphere by the bombardment of cosmic particles from space. As radioactive carbon dioxide it permeates the atmosphere and passes into the bodies of plants and animals through the food chain. To any plant or animal, carbon 14 is indistinguishable from the common carbon (carbon 12) which occurs naturally on Earth. Radiocarbon is relatively rare, so of the total amount of carbon in the body of a plant or animal only a minute fraction is radiocarbon. What makes this tiny fraction useful for dating, argued Libby, is that the *proportion* of radiocarbon is the same for all living animals and plants the world over, and something which can readily be measured.

Radiocarbon begins to decay as soon as it is formed. When a quantity of radiocarbon is produced in the atmosphere, half of that amount will have decayed away (becoming nitrogen gas) in some 5,700 years. Half the remainder will decay in a further 5,700 years, and so on, until an immeasurably small residue remains. Once a plant or animal dies, it ceases to take in radiocarbon from the 'terrestrial reservoir' or outside world, so the amount of radiocarbon in its body begins to dwindle through decay while the ordinary carbon remains unchanged. So, 5,700 years after a tree dies, it contains only half the proportion of radiocarbon to common carbon that exists in a living tree, and in the living world in general. After a total of 11,400 years, or two half-lives, it will contain only one quarter the proportion in the outside world, and so on. After about five half-lives, or roughly 30,000 years, only an immeasurably small residue remains and so the radiocarbon test is only good for dating remains younger than this natural 'ceiling'.

In order to date an organic find (the test only works, of course, on the remains of once-living things, such as bones in a neolithic burial, or Roman fence-posts) it is only necessary to measure the amount of remnant radioactive carbon with a suitable counter and hence deduce when the specimen ceased to take in radiocarbon – when it died.

The great value of the test is that only a tiny fragment of an irreplaceable papyrus or rare skull is needed because it is the proportion of radiocarbon to ordinary carbon that is measured, and compared with the proportions which exist in the terrestrial reservoir or living world today. What the whole technique rests on in the end, therefore, is knowing with some precision just what is the ratio of radiocarbon to common carbon in the terrestrial reservoir today, and it was for making these measurements as well as developing the dating technique that Libby was awarded the Nobel prize.

There is just one further factor of some importance for the test to work properly: the 'standard' mix of radiocarbon to ordinary carbon in the terrestrial reservoir must always have been the same throughout the lifetime of the test subject and in the years since its death. Take the case of archaeologists setting out to determine the age of a neolithic woman whose burial chamber they discover. If there had been a lot more carbon 14 around during the life of this early woman, the reading from her bones will be falsely inflated – she will appear a much more recent burial than she really was. Had there been a lot less radiocarbon around during her life, then the reading will appear falsely diminished and she will appear very much older.

At the time that Libby and his co-workers were developing the new technique, in the 1940s, they had every reason to believe that the amount of carbon 14 in the world could not possibly have varied during the time that mankind had been on Earth simply because the Earth is of immense age, some 4,600 million years old. This great age stamps the radiocarbon technique with the seal of respectability because of what Libby called the 'equilibrium value' for the radiocarbon reservoir.

After the Earth was formed and acquired an atmosphere, there would be a 30,000-year transition period during which carbon 14 would be building up. At the end of that period, the amount of carbon 14 created by cosmic radiation will be balanced by the amount of carbon 14 decaying away to almost zero. To use Libby's terminology, at the end of 30,000 years, the terrestrial radiocarbon reservoir will have reached a steady state.

Since the Earth, according to uniformitarian geology, is many, many times older than the 30,000 years needed to fill up the reservoir, then radiocarbon must unquestionably have attained equilibrium billions of years ago, and must have been constantly so throughout the few million years allotted to human history. To test this essential part of the theory, Libby made measurements of both the rate of formation and the rate of decay of radiocarbon. He found a considerable discrepancy in his measurements; that, apparently, radiocarbon was being created in the atmosphere somewhere around 25 per cent faster than it was becoming extinct. Since this result was inexplicable by any conventional scientific means, Libby put the discrepancy down to experimental error.[5]

During the 1960s, Libby's experiments were repeated by chemists who had been able to refine their techniques after a decade of so of experience. The experiments demand almost heroic measures since the amounts of radiation involved are very small (only a few atomic disintegrations per second) and because of the need to screen out all other sources of radiation which would contaminate the result. The new experiments, though, revealed that the discrepancy observed by Libby was not merely experimental error – it did exist. It was found by Richard Lingenfelter that 'There is strong indication, despite the large errors, that the present natural production rate exceeds the natural decay rate by as much as 25 per cent . . . It appears that equilibrium in the production and decay of carbon 14 may not be maintained in detail.'[6]

Other researchers have confirmed this finding, including Hans Suess of the University of Southern California, writing in the *Journal of Geophysical Research*,[7] and V. R. Switzer writing in *Science*.[8]

Melvin Cook, Professor of Metallurgy at Utah University, has reviewed the data of Suess and Lingenfelter and has reached the conclusion that the present rate of formation of carbon 14 is 18.4 atoms per gram per minute and the rate of decay 13.3 atoms per gram per minute, a ratio indicating that formation exceeds decay by some 38 per cent.[9]

The meaning of this discovery is described as follows by Cook: 'This result has two alternate implications: either the atmosphere is for one reason or another in a transient build up stage as regards Carbon 14 . . . or else something is wrong in one or another of the basic postulates of the radiocarbon dating method.'

Cook has gone one step further by taking the latest measured figures on radiocarbon formation and decay and calculating from them back to the point at which there would have been zero radiocarbon. In doing so, he is in effect using the radiocarbon technique to date the Earth's own atmosphere. And the resulting calculation shows that, using Libby's own data, the age of the atmosphere is around 10,000 years![10]

To anyone who, like me, was brought up on a diet of uniformitarian geology and evolution theory; to any secondary-school pupil or university student who opens a standard geology textbook, the suggestion that life on Earth may have a history as short as 10,000 years inevitably appears preposterous. Surely, the radiocarbon method has been tested against artefacts of known age and has been thoroughly vindicated? Surely the technique has been widely adopted in archaeology with excellent results? And surely any fundamental flaw in the methods would have been discovered years ago?

It is perfectly true that radiocarbon dating has been tried on objects whose age is independently known from archaeological sources and scored some impressive early successes. One of the very first artefacts to be tested was a wooden boat from an Egyptian pharaonic tomb whose age was independently known to be 3,750 years before the present. Radiocarbon assay produced the date of between 3,441 and 3,801 years, a minimum error of only 51 years. But after this promising start, the method quickly ran into difficulties. Anomalous dates were produced from successive assays, including the discovery that some *living* shellfish had very little radiocarbon in their shells and so had been theoretically dead for up to 2,300 years.[11] This made chemists look again at the technique and made them realise that some creatures may interact with parts of the reservoir that are deficient in carbon 14.

In one of the most recent cases of anomalous dating, rock paintings found in the South African bush in 1991 were analysed by Oxford University's radio carbon accelerator unit who dated them as being around 1,200 years old. This finding was significant because it meant the paintings would have been the first bushman painting found in open country. However, publicity of the find attracted the attention of Mrs Joan Ahrens, a

Capetown resident, who recognised the paintings as being produced by her at art classes and later stolen from her garden by vandals. The significance of incidents such as this is that mistakes can be uncovered only in those rare cases where chance grants us some external method of checking the dating technique. Where no such independent verification exists, we simply have to accept the verdict of carbon dating.

The position resulting from these anomalous discoveries was summarised by Hole and Heizer in their *Introduction to Prehistoric Archaeology*:

> For a number of years it was thought that the possible errors . . . were of relatively minor consequence, but more recent intensive research into radiocarbon dates, compared with calendar dates, shows that the natural concentration of carbon 14 in the atmosphere has varied sufficiently to affect dates significantly for certain periods. Because scientists have not been able to predict the amount of variation theoretically, it has been necessary to find a parallel dating method of absolute accuracy to assess the correlation between carbon 14 dates and the calendar.

The parallel dating method turned to in order to assess radiocarbon dating involves that strange tree the bristle cone pine, which grows at high altitudes in the mountains of California and Nevada and is the oldest living thing on Earth – some specimens said to be 5,000 years old.

The bristle cone pine has been exploited by Charles Ferguson of Arizona University to develop the science of dendrochronology – dating by tree rings. The tree is useful here because it lives to a great age and certain 'signature' sequences of tree rings are said to be characteristic of specific years before the present, enabling a younger tree to be correlated with older trees (including dead ones) to stretch the tree-ring chronology further and further back. Cross dating from one core sample to another by means of such signatures enabled Ferguson to construct a master chronology which spans a total of 8,200 years before the present. This has been used to check up on radiocarbon dating variations.

Hans Suess of the University of California in San Diego has radiocarbon dated the bristle cone pine samples of the master chronology and from this a table of deviation has been drawn up which in theory allows the inaccuracies of the radiocarbon method to be corrected for up to around 10,000 years ago.

Radiocarbon dating's inventor Willard Libby did not at first think that large deviations were possible. 'When we developed the radiocarbon dating method,' he said, 'we had no choice than to assume that the cosmic rays had remained constant, though obviously we hadn't the slightest evidence that this was so. But know we know what the variations were.'

Hans Suess was able to show precisely how variations in the amount of cosmic radiation changed the amount of radiocarbon in the atmosphere and his table indicates that by about 5000 BC, radiocarbon-derived dates are around 1,000 years too young.

'Whatever the source of radiocarbon,' says Libby, 'it mixes very rapidly with life on earth so we have a firm belief that the calibrations with the bristle cone pine apply world wide.'

Are archaeologists happy with this result? In fact they appear rather confused by it. Before the bristle cone pine amendments, the dates given by radiocarbon dating had confirmed the widely held belief of diffusionists – that culture had spread from Egypt and the middle east via Mycenae and Crete westward into Europe and then Britain. However, the new chronology indicates that for instance the island of Malta was carving spiral decorations and erecting megalithic structures *before* the supposed cradle civilisations further east. Many archaeologists are unhappy about this but the chronology now has the authority of both Libby and the dendrochronological corrections of Suess's bristle-cone pine deviation tables.

A further difficulty has more recently been introduced into the controversy because the fundamental principal on which dendrochronology is based – that a tree ring forms each year – has been questioned. R. W. Fairbridge, writing on dendrochronology in *Encyclopaedia Britannica*'s entry on the Holocene epoch, says:

> As with Palynology, certain pitfalls have been discovered in tree-ring analysis. Sometimes, as in a very severe season, a growth ring may not form. In certain latitudes, the tree ring's growth correlates with moisture, but in others it may be correlated with temperature. From the climatic viewpoint these two parameters are often inversely related in different regions.[12]

It is also possible for two tree rings to grow in a single year, when growth begins in spring but is later arrested by a period of unseasonal frosts and later still starts again.

These climatic variations presumably mean that a fresh set of correction tables will be needed to modify the bristle cone pine dates, though no one has yet devised a method of calibration for such tables. But whatever the outcome of the debate between archaeologists and radiocarbon chemists, the key question for chemistry is how to explain the observed discrepancy between the rate of production of carbon 14 and its rate of decay in the atmosphere. As Cook has suggested, a possible explanation of the discrepancy is that the atmosphere is still in non-equilibrium because the required 30,000 years have not yet elapsed since it was first formed. We even have a working age of 10,000 years, predicted by the radiocarbon assay equation itself.

Yet how could the Earth possibly be so young? How could science have got things so wrong?

CHAPTER 4

Rock of Ages

In comparison with the 4,600 million years claimed by Darwinists, 10,000 years appears a ludicrously brief time for the age of life on Earth – scarcely more than the extent of recorded history, which stretches back about 7,000 years to the invention of writing in the city of Susa in Mesopotamia.

At the end of the last chapter, I asked: How could science have got things so wrong? The answer turns out to be that it is not science which has got things wrong, merely those scientists seeking to defend a single idea – Darwinian evolution. Science has proposed many methods of geochronometry – measuring the Earth's age – all of which are subject to some uncertainties, for reasons I shall describe in a moment. But of these many methods, one and only one technique, that of the radioactive decay of uranium and related elements, yields an age for the Earth of billions of years. And it is this one method which has been enthusiastically promoted by synthetic evolutionists and uniformitarian geologists, while all other methods have been neglected.

So successful has this promotional campaign been that today almost everyone, including scientists working in other fields, has been led to believe that radioactive dating is the only method of geochronometry worth considering, and that it is well-nigh unassailable because of the universal constancy of radioactive decay.

In fact, none of these widely held beliefs is supported by the evidence. What is more, the dozen or so methods of geochronometry which have been devised in recent years, and which do not depend on radioactive decay, are all in remarkably close agreement about an age for the Earth which is orders of magnitude less than the date derived through radioactive decay measurements.

Details of these alternative methods and their results are described in detail in Chapter 5 but first it is illuminating to look a little more closely at the problems which confront the geologist attempting to measure the Earth's age.

All methods of measuring time, whether for domestic or scientific purposes, rely on the same basic principle: that of monitoring the rate of some constant natural process. Today our most sophisticated chronometric methods involve the rate at which a quartz crystal vibrates when an electric potential is applied to it, and the rate at which radioactive materials decay – said to be the most constant source of all.

But having some readily available process to measure is not enough by itself. To measure elapsed time accurately we must be sure that the process does in fact remain constant, even when we are not watching to check up. You must know the starting value of the clock – how much water was in your water clock to begin with or how tall the candle was before it was lit. And you must be sure that some external factor cannot interfere with the process while it is carrying on: that for instance rainwater does not seep through the roof and top up the water clock, or that a temporary power cut does not stop the electric clock while you are out walking your dog.

All these conditions apply to measuring time today. When it comes to the science of geochronometry, the process we choose will have been going on in prehistoric times which we have no method of directly observing and verifying. This means we must make sure as far as possible that our three conditions were met in the past as well as in the present – and it is here that our problems begin.

Suppose, for instance, we were to take the increasing salinity of the oceans as a means of finding out how old the Earth is (a method actually proposed in 1898 by Irish geologist John Joly). On the face of it this is a promising method since it can be assumed that initially the oceans consisted of fresh water, and the present-day accumulation of salt is due to erosion of land masses by rainfall and the subsequent transport of dissolved salt into the seas by way of the world's rivers. Even more encouraging is the fact that the rate of erosion of the land by rainfall is surprisingly constant each year when totalled up for all continents (about 540 million tons of salt a year). All that would be necessary is to measure the present-day concentration of salt in the sea (32 grams per litre), calculate from this the total amount in all the oceans (about 5×10^{16} tons) and divide this total by the annual amount of salt deposited to get the age of the Earth in years.[1]

Using this method, Joly came up with an age of 100 million years. Unfortunately, when we apply the three conditions mentioned earlier to this method its shortcomings quickly become obvious. We cannot be sure that the annual run-off of dissolved salt has always been constant. Indeed there is good reason to suppose that climatic conditions have been very different in the past – with ice ages and major droughts for instance – and these might have had an effect that is incalculable.

Second, we cannot be quite sure that there was zero salt in the sea to begin with. Conditions might have been sufficiently different initially that some salt might have been present though no one can say how much, if any. (Some recent research in the Atlantic suggests that salts may have been extruded into ocean basins from the molten magma beneath the crust.) And third, it turns out that an apparently constant process is interfered with by external factors. Large amounts of salt are re-circulated into the atmosphere, and recent evidence suggests that the salt in the sea might actually be in a steady state – that as fast as salt is deposited in the sea, it is picked up in the air and re-deposited on land again. A large quantity of salt is evaporated off by biological processes and still more is incorporated into bottom sediments through chemical processes, for instance.

All methods of measuring the age of the Earth are subject to some extent to the same defects – that quite simply, no one was there at the time to check up on our three criteria. The technique used by uniformitarian geologists to arrive at the tremendous age of 4,600 million years for the Earth is usually referred to simply as the 'uranium' or 'uranium-lead' method. Sometimes it is popularly referred to merely as radioactive or radiometric dating. The technique in question covers a family of methods involving the radioactive decay of a number of different metallic elements with very long half-lives (that stay radioactive for very long periods). These elements include uranium and its sister element thorium which both decay into helium and lead;

rubidium which decays into strontium; and potassium which decays into the gas argon.

The basic principle is this: over very long periods of time, uranium spontaneously decays into lead and helium gas. The rate of decay is remarkably constant. The atoms of the uranium are unstable and periodically throw out an alpha particle which is the nucleus of an atom of helium. It is impossible to tell in advance when any particular atom will break apart in this way since the process occurs at random. But in any substantial mass of the mineral, there will be many billions of atoms and, with very large numbers of events, the 'law of large numbers' operates to produce a statistically predictable result.

The important part of the theory is that the kind of lead into which uranium eventually decays is chemically distinct from common lead already present in the rocks, and is referred to as radiogenic lead; a daughter product of the decay process. Common lead is an isotope called lead 204, while the decay product of uranium 238 is lead 206. In order to date a rock deposit, a sample is taken and the amount of radioactive uranium, together with the amount of radiogenic lead it contains is accurately assayed in the laboratory. Since the rate of decay is known from modern measurements, it is possible to calculate directly how long the uranium has been decaying – how old the deposit is – by how much radiogenic lead it has turned into.

The half-life or decay constant of uranium 238 (one of the principal isotopes used) has been calculated to be 4,500 million years. To take a simplistic example, if the assay showed that a deposit was composed of half uranium 238 and half lead 206, then one would draw the conclusion that the deposit was 4,500 million years old. (This, incidentally, is the average figure that *is* found for the Earth's crust although the figure is arrived at by extrapolation rather than direct measurement.)

On the face of it, uranium decay seems the ideal method of geochronometry, and scientifically above suspicion. But, as in the case of radiocarbon dating, research in recent decades has begun to cast serious doubts on its reliability.

First there is the rate of decay itself. At present this process is perfectly constant the world over, a fact amply confirmed by repeated measurement. When we apply the technique to dating the past, though, we come up against the requirement that the process must always have been constant – and here lies the difficulty. As long as the radioactive decay of a uranium ore deposit is left undisturbed it does remain constant. But when certain external factors intervene, our atomic clock can be speeded up considerably. This fact, of course, is the very basis of the nuclear industry where the uranium is artificially enriched (by concentration) in order to speed up the decay process and thus generate controlled amounts of heat for use in power stations. Enrichment can also occur naturally by the redistribution of soluble uranium ores and there is evidence of a fossil nuclear reactor in Gabon in Africa; evidence that spontaneous nuclear burning has taken place in uranium ore deposits – fortunately for us in prehistoric times – probably as a result of such natural enrichment.[2]

But as well as enrichment of the uranium, there are other factors which may also speed up the decay process. One of these is the concentration of highly energetic neutrino particles caused by cosmic radiation from outside the Earth. Usually cosmic rays, and the particles resulting from cosmic ray collisions, are stopped by the atmosphere before they reach the Earth's surface. But occasionally, some more than usually energetic particles get through and bombard the rocks of the crust, often penetrating deeply underground. Periods of sunspot activity can have this effect for example.

It is now believed there are two phenomena which would have the effect of producing showers of neutrinos that would greatly increase the rate of radioactive decay in the rocks they penetrated: a supernova explosion of a nearby star, or a reversal of the Earth's magnetic field – and both these phenomena are known to have taken place, not once but many times.

According to Frederick Jueneman, writing on supernovae in *Industrial Research,*

> Being so close, the anisotropic neutrino flux of the super-explosion must have had the peculiar characteristic of resetting all our atomic clocks. This would knock our carbon-14, potassium-argon and uranium-lead dating measurements into a cocked hat. The age of prehistoric artefacts, the age of the Earth, and that of the universe would be thrown into doubt.[3]

The Crab Nebula, a feature of the night sky familiar to astronomers, is the remains of a relatively nearby star which exploded and was observed by Chinese astronomers on 4 July

1054. The supernova was visible in daylight for twenty-three days and at night for nearly two years. At least four supernovae have occurred within the Earth's galaxy in historical times and over 230 have been mapped in the sky since 1885. The supernova in the constellation Lupus, observed in 1006, is believed to have been the brightest and the nearest (3,000 light years away).

On 24 February 1987 a supernova occurred in the Large Magellanic Cloud and produced a neutrino shower that was detected at several laboratories on Earth. A team of Italian and Russian scientists working near Mont Blanc reported a burst of neutrinos on the morning of 23 February. Separate bursts of neutrinos were detected by laboratories in Kamioka, Japan, and Ohio, USA, at about the same time. Particles recorded by the American team had energies of between 30 million and 100 million electron-volts.

Writing on palaeomagnetism in the 1984 edition of *Encyclopaedia Britannica*, J. R. Heirtzler says:

> An interesting aspect of the reversal chronology observed in deep sea cores is the apparent correlation of the extinction and first appearance of certain marine fossil organisms with reversal boundaries. It has been speculated that this is a causal relationship resulting from an increased influx of cosmic radiation during a polarity reversal.

That the Earth's magnetic field has reversed itself (the north magnetic pole becoming south, and vice versa) is known from remnant magnetism in primary or igneous rocks such as those thrown up from volcanoes or in sea-bed lava flows through cracks in the ocean floor. When rock is cooling from a liquid state, any iron particles it contains will align themselves with the Earth's magnetic field as the rock solidifies and become trapped as a sort of natural tape recording of the direction of the field, 'frozen' in the rock. Surveys of such palaeomagnetism have disclosed that the magnetic field has reversed not once but many times, leaving alternating magnetic 'stripes' frozen as a permanent record in the continuously extruded once-molten rock as it spreads over the sea floor (see, for example, N. D. Opdyke writing on 'Palaeomagnetism of Oceanic Cores' in R. A. Phinney's *History of the Earth's Crust* (1968), and F. J. Vine on 'Magnetic Anomalies Associated with Mid-Ocean Ridges' in the same volume).

It is not possible to say at present how far these phenomena will have speeded up radioactive decay: how far they will have re-set our atomic clocks. But it is significant that the speeding up of the decay process would have had the effect of *lengthening* the time-scales measured with their aid; of making the Earth appear older than it actually is. This effect, incidentally, would be felt not merely by uranium dating but all radioactive dating techniques, including radiocarbon dating.

The second criterion for any method of geochronometry is that we must know the starting value of the process we are measuring; we must have a point of departure or reference point from which to make our calculations. Again, on the face of it, uranium decay fulfils this requirement since the type of lead which results is said to be uniquely formed as a by-product of this process. If radiogenic lead – lead 206 and lead 207 from uranium, and lead 208 from thorium – really is uniquely formed as the end product of disintegration, then it is perfectly reasonable to suppose – as adherents of radioactive dating do – that there was zero radiogenic lead in the rocks of the Earth's crust when they first formed, and so we have a reliable starting point for our calculations. The same argument can be used to make us reasonably certain that no radiogenic lead could have got into the rocks by some other means, thus distorting the effects of the decay process.

But once again things are by no means as simple as they seem when investigated a little more closely. For it appears that there is another and quite separate mechanism by which common lead can be transmuted into a form which on assay will be indistinguishable from 'radiogenic' lead. This transmutation can occur through the capture of free neutrons – atomic particles with enough energy to transmute common lead into so-called radiogenic lead. Where, though, could such a source of free neutrons be found? The answer is in a radioactive ore deposit such as uranium, where they occur spontaneously through radioactive decay! In other words, the very process being measured can be, as it were, moonlighting at another job. As well as spontaneously decaying into radiogenic lead, it is also making available a supply of particles which are simultaneously converting common lead into another isotope which, on being assayed, will be indistinguishable from a radiogenic product of alpha decay. Significantly, here too we have a mechanism that would weight our measurements in favour of an 'old' Earth, for too much 'radiogenic' lead would lead us to imagine that the process had been going on for much longer than it actually has.

According to Henry Morris,

> An even more important phenomenon . . . is that of 'free neutron capture', by which free neutrons in the mineral's environment may be captured by the lead in the system to change the isotopic value of the lead. That is, lead 206 may be converted into lead 207, and lead 207 into lead 208 by this process. It is perhaps significant that lead 208 usually constitutes over half the lead present in any given lead deposit. Thus the relative amounts of these 'radiogenic' isotopes of lead in the system may not be a function of their decay from Thorium and uranium at all, but rather a function of the amount of free neutrons in the environment.[4]

In his *Prehistory and Earth Models*, Cook has analysed the lead content of two of the world's largest uranium ore deposits – in Katanga and Canada. He found that they contained no lead 204 (no lead from non-radiogenic sources) and practically no thorium 232. However they do contain significant amounts of lead 208. This could have been derived only from lead 207 by neutron capture, says Cook, while all the so-called radiogenic lead can be accounted for on the same basis and the mineral deposits could be essentially of modern origin.[5]

So uranium decay fails our second criterion as well. But it also fails our third – that we must be reasonably sure no outside agency can interfere with the smooth running of our chosen process. Uranium does not naturally occur in metallic form but as uranium oxide. This material is highly soluble in water and is known to be moved away from its original deposit in large quantities by ground waters. The type of effect this has on dating is unpredictable since some parts of a mineral deposit can be unnaturally enriched while others are unnaturally depleted. So viewed overall, uranium-lead decay fails to fulfil any of the criteria required of a reliable method of geochronometry, and yet it continues to be enthusiastically supported by uniformitarian geologists and Darwinian evolutionists.

There is one further discovery relating to uranium dating which is of considerable relevance to the age of the Earth. As mentioned earlier, the final disintegration products of the decay process are two: not only lead but also helium gas. Like the lead which results from the decay process the helium is also a radiogenic daughter product, with an atomic weight of 4. In fact almost the entire amount of helium in the Earth's atmosphere is believed to be radiogenic helium, formed during the decay process throughout the major part of the Earth's history.

Now, if the uranium-lead dating technique were reliable, then the amount of this radiogenic helium in the atmosphere would yield a date for the Earth's age consonant with that yielded by measuring the amount of radiogenic lead in the crust. In fact, the dates are so different as to be irreconcilable.

If the Earth were 4,600 million years old, then there would be roughly 10,000 billion tons of radiogenic helium 4 in the atmosphere. Actually, there is only around 3.5 billion tons present – several thousand times less than there should be (0.035 per cent to be precise).

Writing in *Nature* on the 'mystery' of the Earth's missing radiogenic helium, Melvin Cook says:

> At the estimated 2×10^{20} gm uranium and 5×10^{20} gm thorium in the lithosphere, helium should be generated radiogenically at a rate of about 3×10^9 gm/yr. Moreover the (secondary) cosmic-ray source of helium has been estimated to be of comparable magnitude. Apparently nearly all the helium from sedimentary rocks and, according to Keevil and Hurley, about 0.8 of the radiogenic helium from igneous rocks, has been released into the atmosphere during geological times (currently taken to be about 5×10^9 yr). Hence more than 10^{20} gm of helium should have passed into the atmosphere since the 'beginning'. Because the atmosphere contains only 3.5×10^{15} gm helium 4, the common assumption is therefore that about 10^{20} gm of helium 4 must also have passed out through the exosphere, and that the present rate of loss through the atmosphere balances the rate of exudation from the lithosphere.[6]

As Cook says, uniformitarian geologists have attempted to explain this discrepancy by assuming that the other 99.96 per cent has escaped from the Earth's gravitational field into space – but this process has not been observed. On the contrary, recent thinking has suggested that far from losing radiogenic helium, the atmosphere may actually be gaining quantities of this material. As it orbits the Sun, the Earth moves through a thin solar atmosphere, which consists principally of hydrogen and helium resulting from nuclear processes within the Sun, and measurements in the upper atmosphere have suggested that the Earth is gaining helium by this means.

In his book *Gaia: A New Look at Life on Earth*, space scientist James Lovelock writes:

> The outermost layer of the air, so thin as to contain only a few hundred atoms per cubic centimetre, the exosphere, can be thought of as merging into the equally thin outer atmosphere of the sun. It used to be assumed that the escape of hydrogen atoms from the exosphere gave the Earth its oxygen atmosphere. Not only do we now doubt that this process is on a sufficient scale to account for oxygen, but we rather suspect that the loss of hydrogen atoms is offset or even counterbalanced by the flux of hydrogen from the sun.[7]

Of course, Lovelock is writing of hydrogen not helium. However, helium is four times heavier than hydrogen and it is extremely plentiful in the sun's atmosphere since it is the principal product of the sun's nuclear fusion process. If hydrogen is not lost but gained, then the same will be true for helium.

If we take the measured amount of helium 4 in the atmosphere and apply the radioactive dating technique to it, says Cook, we find that the calculation yields an age for the Earth of *around 175,000 years*. This procedure fails our criteria of reliability on two counts: that uranium decay has probably been accelerated by an unknown amount in the past, and that the possible acquisition of helium 4 from outside upsets the process. But both these factors indicate that the figure of 175,000 years is inaccurate in that it is too great.

What about the dating techniques based on other radioactive elements referred to earlier? Morris points out that the methods based on the decay of potassium to argon and rubidium to strontium are also subject to the defects already described, as well as having specific problems of their own:

> The method most widely used for dating rocks is the potassium-argon method. Potassium minerals are found in most igneous and some sedimentary rocks . . . Potassium 40 decays by the 'electron capture' process (capture of an orbital electron by the nucleus) into argon 40, with a half life of 1.3 billion years. It also

decays simultaneously by the 'beta-decay' process (emission of an electron and a neutrino) into calcium 40.[8]

The process is complicated by the fact that the potassium simultaneously decays into two different radiogenic daughter products at two different rates, one roughly eight times faster than the other, giving a differential decay rate or 'branching ratio' which has defied accurate measurement. In practice this means that the potassium-argon method has to be 'calibrated' by the uranium-lead method and hence cannot be looked to to provide any greater accuracy. Indeed, the branching ratio is commonly 'adjusted' in order to improve the concordance between potassium-argon dates and uranium-lead dates.

The potassium-argon method is further complicated because the end products, argon 40 and calcium 40, are very common isotopes in both the atmosphere and the rocks of the Earth's crust. Cook has calculated that no more than 1 per cent of the argon 40 currently present on Earth could be a radiogenic daughter product and it is highly probable that some of the argon 40 in all potassium minerals has been derived directly rather than as a result of decay.[9]

As pointed out by Morris, this probability is borne out by numerous studies of volcanic rocks which have resulted in anomalous dating, even to the extent that modern volcanic lavas formed in recent historical times have been dated as up to 3 billion years old by the potassium-argon method.

According to Noble and Naughton of the Hawaiian Institute of Geophysics:

> The radiogenic argon and helium contents of three basalts erupted into the deep ocean from an active volcano (Kilauea) have been measured. Ages calculated from these measurements increase with sample depth up to 22 million years for lavas deduced to be recent. Caution is urged in applying dates from deep-ocean basalts in studies on ocean-floor spreading.[10]

Morris quotes a similar study of Hawaiian basaltic lavas actually dating from an eruption in 1801, near Hualalei, which were found to give potassium-argon dates ranging from 160 million years to 3 billion years.[11] The suspected reason for the anomalous ages was the incorporation of environmental argon 40 at the time of the eruption.

Radiogenic strontium – strontium 87 – occurs in rocks as a result of decay of radioactive rubidium. However, the half-life or time constant of this decay process is not known with any accuracy. Determinations made in the laboratory have produced figures varying from 47 billion years to 120 billion years. The average figure obtained by direct measurement of 68 billion years yields ages for rocks as much as 45 per cent different from measurement of the same rocks using the uranium-lead method. Another problem is that strontium 87 occurs both as a radiogenic daughter product of radioactive decay *and* as a commonly occurring element in its own right. Typically, rocks contain ten times more common strontium 87 than radiogenic strontium 87. Rubidium-strontium is also suspect because it is subject to exactly the same neutron capture process as uranium-lead. This time it is strontium 86 which can be transformed to strontium 87.

Most disconcerting of all is the fact that these various methods of dating commonly produce discordant ages for the same rock deposit. Where this occurs, a 'harmonisation' of discordant dates is carried out – in other words, the figures are adjusted until they seem right.

In one sense it is rather surprising that Darwinists have embraced radioactive dating techniques so enthusiastically. True, these techniques have provided the billions of years of pre-history required by the theory. But – almost incredibly – radioactive dating is otherwise quite useless to evolutionists since it cannot be used directly to date sedimentary rocks – the rocks in which fossils are found. That this is so will astonish many teachers of geography and geology who have been led to believe that radioactive dating is the ultimate authority for the assigned ages of rock strata and the fossils contained in them.

What this means is that the majority of the superficial features of the Earth, the majority of the sedimentary rocks which cover large portions of every continent, and the fossil record of earlier life they contain are not directly accessible to radioactive dating. In Britain, for instance, more than three-quarters of the country is covered by sedimentary rocks: only in relatively small areas of Cornwall, Wales and the Scottish Highlands are volcanic or igneous primary rocks exposed. Dartmoor is a typical granite outcrop. These rocks can be used for radioactive dating (with all its flaws) but they contain no fossils. The same is true of the inhabited land masses on the five continents.

The entire geological column of sedimentary rock strata which is a prominent feature of every natural history museum, annotated with the age of each deposit and the type of animal or plant life dominating each stratum, is dated not by radioactive decay or any other method of direct measurement, but by relative methods. These methods include principally the dating of intrusions of primary volcanic rock and ash (often by the potassium-argon method of Hawaiian fame) and estimates based on assumptions concerning rates of sedimentation and rates of evolution of fossil species. When the British Museum says, for instance, that the Devonian period began around 400 million years ago and ended some 350 million years ago, these dates are not based on determinations made directly by radioactive decay because rocks of Devonian age contain only sediments composed of minerals of uniform earlier age. The age of 400 million years is an estimate based on factors such as how long it would take for the succeeding sediments to be laid down assuming ocean-bottom conditions comparable with those of today; and on the evolutionists' estimate of how long it would take for life to evolve from the state represented by Devonian fossils to its present-day state.

The fact is that it is impossible to say at present with any confidence how old the Earth is, beyond the fact that it predates the calendar of human history. However, there is a range of indicators available that are worth considering in more detail.

CHAPTER 5

Footprints in the Moon Dust

In the preceding chapters, I presented evidence of the unreliability of methods of geochronometry based on radioactive dating techniques. Although they receive little or no attention in school curricula or textbooks, there are a dozen methods of geochronometry other than radioactive decay – some of which have been available for many decades. Although varying considerably in their approach to the problem and the kind of natural processes they employ, all are remarkably similar in pointing to a relatively young Earth: an age measured in millions (or even hundreds of thousands) rather than in billions of years.

The following chapter is a summary of the main methods: meteoric dust; persistence of interplanetary dust; persistence of short-period comets; magnetic field decay; dissolved nickel in the oceans and alternative radioactive methods. However, I feel compelled to admit that it is rather dry technical stuff. Readers who prefer not to wade through such details may feel happier skipping to the end of the chapter, where the ages yielded by these methods are presented in a summary table.

METEORIC DUST

Cosmic dust particles and micrometeorites continually enter the Earth's atmosphere from space and settle on the surface. Hans Pettersson of the Oceanographic Institute of Goteborg has measured the rate at which this dust arrives and has found that it is constant from year to year at roughly 14 million tons annually.[1]

The implications of this finding are pretty clear. We are dealing with a process which consists of an aggregation of random microevents; which is not known to be interfered with by any outside agency; and whose starting value in the present context is not relevant. It is not relevant because it is the *absence* of dust we are concerned with. Morris points out that if the Earth is 4,500 million years old, then some 63 million billion tons of dust have settled on its surface. The Earth has a surface area in the region of 5.5×10^{15} square feet, and compacted dust can be assumed to have density in the region of 140 pounds per cubic foot. These figures indicate that enough meteoric dust would have entered the atmosphere to create a layer 180-feet thick. This layer clearly does not exist, nor anything like it.

It has been suggested that all or most of this dust has fallen into the ocean, onto the sea bottom, and any remainder has become mixed with terrestrial dust from the rocks of the Earth's crust. But the suggestion in unconvincing since the mineral composition of meteorites and meteoric dust is chemically distinct from the rocks of the Earth's crust, containing large amounts of nickel and iron which are both relatively rare elements in the Earth's crust. Pettersson estimated the average nickel content of meteoric dust to be 2.5 per cent, some 300 times greater than that in the rocks of the crust. There are no concentrations of nickel such as this either on the Earth or on the ocean bottoms. According to Morris, there is a total of around 7,000 billion pounds of nickel dissolved in the oceans. The total amount of nickel carried from the land mass into the seas each year is around 750 million pounds. This means that, on one hand the total nickel dissolved in the seas could have been transported there in as little as 9,000 years, and on the other, there does not appear to be a hidden reservoir of meteoric nickel.[2]

The dust problem applies not only to the Earth, but to the Moon as well, which uniformitarians also believe to be billions of years old. Before the first manned landing by Apollo 11 in 1969, it was feared by some lunar geologists that the dust layer on the Moon's surface – undisturbed by atmospheric movement or by oceans – might be so thick that the landing craft would simply disappear into a sea of dust. In the event, of course, Neil

Armstrong took his 'giant leap' into merely an inch or two of dust – the sort of amount that would accumulate in thousands, rather than billions of years.

ASTRONOMICAL STUDIES

Astronomical measurement and observation and the use of computer simulation to study the formation of the solar system have suggested two methods of indicating an upper limit to the Earth's age. First there are the short-period comets – such as Halley's comet, Arund-Rolenson, and the recently discovered Kohoutek. Russian astronomer Professor S. K. Vsekhsviatsky, Director of the Kiev Observatory, has studied periodic comets extensively and written two standard works on the subject. He has come to the conclusion that they are losing their luminosity and the matter which constitutes them at such a rapid rate that a comet will disintegrate completely within 50 to 60 revolutions of the solar system.[3] Halley's comet could thus be less than 6,000 years old.

Several short-period comets observed over the last 100 years have failed to return and there are cases where comets appear to have disintegrated while being observed, confirming Vsekhsviatsky's prediction. This has happened very recently, in May 1991, when Halley's comet, after its latest rendezvous with Earth, was seen to break up as it moved away.

In addition there is the Poynting-Robertson effect which predicts that interplanetary dust particles will be pushed out of the solar system and into space by the pressure of radiation from the Sun, while larger particles will be slowed down in their orbits and swept into the Sun. The existence of a solar wind capable of causing such effects was predicted in 1951 and confirmed experimentally in 1962 by a US spacecraft sent to Venus. The probe found a steady stream of protons and electrons around the earth with a density of between one and ten particles per cubic centimetre. Significantly, both of the dust dispersion effects are predicted to take place in less than 100,000 years after planetary formation.

DECAY OF THE EARTH'S MAGNETIC FIELD

Dr Thomas Barnes, Professor of Physics at the University of Texas, has made a special study of the Earth's magnetic field. Morris points out that the strength of the magnetic field has been measured scientifically since the ninteenth century and has shown from statistical studies of these measurements that the field is decaying exponentially with a half-life he calculates to be around 1400 years.[4]

This rate of decay is relevant to measurements of the Earth's age, because it is a pointer to a practical upper limit. The half-life value calculated for the magnetic field means that 1,400 years ago the flux density of the field (its 'magnetic strength') was twice as strong as today; 2,800 years ago it was four times as strong and so on. Some 7,000 years ago the field would have been 32 times as strong as at present, and 10,000 years ago the field would have been thousands of times its present strength.

While this measuring technique does not make possible a direct measurement of the Earth's age, it places a limit on the possible age range which may be considered for life on Earth and limits that range to something in the order of 10,000 years or less. Beyond this the flux density of the Earth's magnetic field would have been impossibly great.

Finally, there are the short timescale indicators referred to in the last two chapters.

RADIOGENIC HELIUM

As observed earlier, one of the defects of the uranium-lead-derived age of the Earth is that it is not supported by the age given by the other product of the self-same decay process, radiogenic helium. It has been estimated that in the rocks of the Earth's crust there is roughly 40,000 billion tons of uranium and roughly 100,000 billion tons of thorium. As it decays radioactively, this mass of rock will generate roughly 3,000 tons of radiogenic helium each year which will enter the atmosphere. According to Keevil and Hurley, practically all the radiogenic helium from sedimentary rocks has been expelled into the air during the geological past, together with around 80 per cent of that from igneous or primary rocks. If, as evolutionists believe, the Earth is 4.5 billion years old, there should be around 10,000 billion tons of radiogenic helium in the atmosphere. In fact there is only 3.5 billion tons present – 0.035 per cent of the expected figure.[5]

If we take the measured amount of radiogenic helium at its face value, and make the conservative assumption that the atmosphere, to begin with, contained no radiogenic helium, then the Earth's age comes out at about 175,000 years. The major sources of uncertainty about this dating method are, first, that there may have been some helium 4 present to begin with; second that radiogenic helium may well be entering the atmosphere from the Sun; and third that alpha decay may have been speeded up by outside factors. In each case the figure of 175,000 years is inaccurate in that it would be too great.

NON-EQUILIBRIUM OF CARBON 14

Earlier, in Chapter 3, I quoted the measured rate of generation of carbon 14 in the atmosphere as exceeding the rate of extinction by as much as 38 per cent, indicating that the 30,000 years or so needed by the terrestrial reservoir to attain equilibrium have not yet elapsed. Using Libby's own equations, the atmosphere can be 'dated' by carbon 14 as being roughly 10,000 years old.

CONTINENTAL DRIFT

As described in the next chapter, it is now generally accepted that the present continental land masses are fragments of a prehistoric super-continent. Although uniformitarians have attempted to reconcile this break-up with long time scales and slow gradual processes, the only realistic model so far proposed is the ice-cap rupture model of Cook, which entails sudden explosive processes and which involves an ancient Arctic being dissipated as recently as 10,000 years ago.

Once again, this does not rule out an ancient Earth, but it does place severe limitations on the antiquity of origin of many species which is at present explained by geographical isolation, as explained in detail in the next chapter.

The geochronometries and time scale indicators described above are summarised in the table below. Looking through the body of evidence they represent, I cannot help wondering: if it were not for the demands of Darwinian evolution, for whose random genetic mutations vast reaches of time are indispensable, how many rational geologists would continue to believe that the Earth must be billions of years old? That a young Earth is impossible?

TABLE 1

Short Time-Scale Geochronometries and Indicators

METHOD	INDICATED AGE OF EARTH
Radiogenic helium in atmosphere	Less than 175,000 years
Poynting-Robertson effect	Less than 100,000 years
Persistence of interplanetary dust	Less than 100,000 years
Non-equilibrium of carbon 14	Less than 30,000 years
Persistence of short-period comets	Less than 10,000 years
Magnetic field decay	Less than 10,000 years
Dissolved nickel in oceans	Less than 9,000 years
Meteoric dust in atmosphere	Recent origin of Earth
Continental drift (ice-cap rupture)	Recent origin of life

PART TWO

Clay

CHAPTER 6

Tales from Before the Flood

In 1922, archaeologist Sir Leonard Woolley began to excavate the remains of one of the world's oldest cities, between the rivers Tigris and Euphrates in Mesopotamia, or present-day Iraq. Woolley's hopes of great discoveries at the site of the biblical city of Ur were more than fulfilled. But what he found not only caught the archaeological world by surprise, it also sent a ripple of consternation spreading through the world's natural history and geological museums.

Six thousand years ago, civilisation arose in the plains of Sumeria, where many famous cities flourished and died. It was here that the legendary kings of Babylon lived and here that writing was invented. The fame of Ur has outlasted many another Sumerian city because the Bible gives it as the birth-place of the patriarch Abraham – 'Ur of the Chaldees'.

The site of the city was identified at the end of the last century as present-day Tell al Muqayyar ('the mound of pitch') through clay cylinders inscribed in the Akkadian language. Woolley's expedition was sent out by the British Museum after the First World War to examine and report on the remains.[1]

In common with other cities of the Sumerian plain, Ur today is little more than a gigantic mound of rubble: ruin piled on ruin as each generation simply constructed new houses and public buildings directly on top of old ones which had crumbled with age, or which were knocked down to make way for newer property developments. By cutting trenches straight down through the mound, Woolley planned to reveal a slice of the history of Ur, its people and their artefacts. Eventually, his excavation took him so deep he found material from a period of immense antiquity which actually predated the Sumerian people and which he named the 'al 'Ubaid period'.

Driving deeper still, Woolley hit what most of his workers took to be the end of their dig, a thick bed of clay and silt. But continuing to dig, he passed through the thick bed of water-laid sediments and emerged again into the remains of civilised life, including 'al 'Ubaid' pottery.

He had clearly found the remains of a great flood. 'No other agency could possibly account for it,' wrote Woolley. 'Inundations are of normal occurrence in lower Mesopotamia, but no ordinary rising of the rivers would leave behind it anything approaching the bulk of this clay bank: eight feet of sediment imply a very great depth of water, and the flood which deposited it must have been of a magnitude unparalleled in local history.' Woolley believed he had found evidence of the great flood of Noah, described in the Bible, and the evidence for this is compelling.

The flood sediments he discovered date from around 3000 BC, early in the establishment of civilisation in the area. Sumerian clay tablets from around 2000 BC give an account of the flood as being divine retribution from a Sumerian god. The deity, however, takes pity on one man, Uta-Napishtim, and tells him to construct a boat, making it watertight with pitch. Uta-Napishtim saves his family and many animals aboard his boat which survives seven days of rain. At the end of this week, Uta-Napishtim sends out a dove and a swallow, which return to the boat. Later he sends out a raven which does not return because it has found dry land. The boat comes to land on a mountain top.

It is widely held today that the Noachian flood story in the the Hebrew bible was borrowed by them from their neighbours the Sumerians. This idea is supported by the fact that while there is ample evidence of such a flood in Sumeria, none has been found in the lands occupied by the men who wrote the Hebrew bible.

The extent of the Sumerian flood was very substantial: a deposit 8 feet thick, covering an area some 400 miles long by 100 miles wide – a total of many billions of tons of material. And it was this discovery that sent a buzz through the corridors of uniformitarian geology. For here at last was evidence of a real *Homo diluvii testis* – man a witness to the Flood. Because this catastrophic event had occurred within recorded history then – uniquely in the geological record – here was direct evidence of a substantial sediment that must have been laid down rapidly and all at once, rather than slowly over millions of years. And if this stratum, then why not others? If, as Hutton believed, the present is the key to the past, then was not this sediment the key to previous rocks formations?

At almost the same time that work began on the excavation of the biblical Ur, a German meteorologist, Alfred Wegener, published a theory that was greeted with universal derision by the world scientific community: the theory of continental drift.[2] Wegener's idea – that the major land masses were once joined, but have subsequently been forced apart – was regarded by geologists little more than forty years ago as belonging to the lunatic fringe of pseudo-scientific beliefs. Yet since the 1960s, the evidence for continental drift has become overwhelmingly convincing and few today doubt its validity.[3] Perhaps to disguise their embarrassment at rejecting the idea so scornfully in the past, uniformitarian geologists have made continental drift respectable by re-christening it as 'plate tectonics' – a subject now included in all geological curricula and textbooks.

In its present-day form, the idea is that the continents are the visible portions of gigantic 'plates' whose edges are for the most part concealed beneath the oceans or deep in the Earth's crust, and which are 'floating' on the semi-fluid material of the Earth's mantle. The continents are thus rather like pieces of cracked egg-shell, floating on a soft-boiled egg.

The reason that continental drift was so disreputable to early twentieth-century geologists, was that the forces required to crack the Earth's crust apart must have been cataclysmic, and this awakened the suspicion of Darwinists that catastrophism was not only about to rear its ugly, dinosaur-like head once more but was actually to gain admittance to the scientific drawing-room through the back-door.

The chief evidence for continental drift is the complementary shape of geological features and coastal outlines of the continents; the apparent wandering of the magnetic poles along different paths for different continents and the young ages of marine sediments and the ocean floors. When allowance is made for their continental shelves, the Atlantic shorelines of Africa and South America appear to be pieces of a former whole

as do those of Europe and North America. Studies of palaeomagnetism show that at some time in the remote past the rocks
of the crust 'pointed' to a different North Pole and have since
moved (although some geologists remain sceptical on this
issue). Drilling near ocean ridges in the Atlantic shows that the
sediments overlying the bedrock get older as you move away
from the ridge crest from where it is thought a continental plate
is spreading.

It is now widely accepted that the land masses that have been
pushed as much as a thousand miles apart probably did form a
single land mass, which Wegener christened 'Pangaea' (all-
Earth). The major (northern) portion of this land mass is usually
today called Gondwanaland after rock strata in India which can
also be traced in South Africa and South America.

The important question for geology, of course, is, just what
agency caused the original continent to break apart? An equally
important question from the standpoint of evolution theory is,
precisely when did this event take place? While there has as yet
been no agreement on the cause of the event, uniformitarians
have predictably dated its occurrence as during the Mesozoic
era, which they believe lasted from 250 million years ago to 65
million years ago.

The central problem with continental drift, or plate tectonics,
and the factor that delayed its acceptance for decades is that no
one has so far proposed a satisfactory mechanism to drive the
process. A number of explanations of the cause of continental
drift have been proposed, each with its merits and difficulties.
They include: tidal forces; expansion of the Earth; convection
currents in the semi-fluid mantle; and successive loading and
unloading of the crust (by glaciers, for instance).

A satisfactory model of the process has to meet three main
criteria. First it must demonstrate a mechanism which can
produce sufficient force to initiate the breaking apart of the
crust. This requires very large amounts of energy to be released,
whether the fracture is caused by compression (like squeezing
an egg in your hand) or by tension (like holding a telephone
directory by its edges and pulling it apart). Second, the proposed mechanism must also provide sufficient energy to drive
the fractured 'plates' apart, in some cases riding over neighbouring plates, and in other cases pushing directly against the
edges of adjacent plates, folding them to form mountain ranges

and, perhaps, thickening the crust in places. This is an important consideration because this 'shouldering aside' of whole continents and mountain-building activity required even more energy than cracking the crust apart in the first place. Finally, the proposed mechanism must provide sufficient energy to break the crust and part the continents, but at the same time not generate excessive amounts of heat. Although it was once believed that the Earth was cooling as its molten interior lost heat, it is now known that the Earth's overall temperature is roughly constant, since heat loss from the surface is balanced by heat generated within the crust by radioactive decay. The continental drift mechanism must not disturb this heat balance.

Looked at from this point of view, most of the mechanisms proposed fail to account satisfactorily for the distribution of land masses that we observe today. Melvin Cook has made a detailed study of the most promising models and has shown that they are incapable of providing the required energies.

Probably the hypothesis most favoured today by uniformitarian geologists is that continental drift is due to mantle convection currents. Underneath the crust, the Earth's mantle is subjected to intense heat and pressure and under these conditions behaves like a semi-fluid. A rough analogy is the semi-molten iron poured from a blast-furnace, with a crust of solid slag floating on top. Heat currents rise through the mantle from the core, travel for some distance along the base of the crust losing heat, and then descend, causing a vast circular movement of the mantle in the vertical plane.

The heated mantle material in one circular current, it is hypothesised, may cause friction against the base of the crust, dragging it apart from an adjacent section of crust, which in turn is being dragged in a different direction by an adjacent mantle convection current.

Cook has calculated that the amount of heat generated by mantle convection great enough to cause continental drift would be between 1,000 and 10 billion times greater than the rate of radiogenic heat generation in the crust as a whole. 'Clearly,' says Cook, 'such currents are impossible because either they would melt the Earth in a very short time, or one would observe an enormously greater heat flux from the Earth than is actually observed.'[4]

There are other objections to convection currents, too. The

measured rate of flow is far below the velocity required by theory for them to be capable of breaking and shifting the continents. In addition, the 'velocity-gradient' of the viscosity of these currents (how runny the semi-fluid material is) would have to be at least 100 million times greater than currently postulated in order to cause continental drift.

Another theory enjoying some popularity in recent years is that the Earth may have expanded, thus cracking the original Pangaea apart. An advantage of this model is that it would also explain the observed expansion of the ocean basins. Although at first sight the idea of an expanding Earth seems rather far-fetched, the theory does have considerable merit. An expansion in the surface area of the Earth of about 45 per cent would account for the separation of Pangaea into today's fragments. To get a 45 per cent increase in surface area would mean an increase in the Earth's diameter of about 20 per cent.

Naturally, true to uniformitarian principles, this expansion is deemed to have taken place over immense reaches of time, during the 250 million years that are said to have elapsed since the Palaeozoic era. This would mean an increase of around one centimetre a year in the Earth's diameter – on the face of it, not an unreasonable amount.

Unfortunately, the problem with this idea arises from exactly the same defect – its energy requirement. To fuel the expansion would take in total the entire chemical energy locked up in bonding together all the matter comprising the Earth. Chemical changes can thus be ruled out as being responsible for terrestrial expansion. The expanding Earth idea lacks a mechanism. In very recent times, however, a further attempt to rescue this idea has been made by postulating the steady-state creation of matter in the Earth's core as fuelling the expansion. The idea that hydrogen atoms might be continually coming into existence naturally was first made popular by astronomer Sir Fred Hoyle who suggested the process might be occurring in the space between the stars. The difficulty here is that the idea is confined wholly to the realm of theory since no one has ever observed or measured the steady-state creation of matter either in the Earth's core or in interstellar space.

Perhaps one of the main stumbling blocks to geologists in arriving at a satisfactory mechanism of the break-up of Pangaea and its re-distribution around the Earth has been the persistence of the idea of continental 'drift'. Loyally obeying uniformitarian principles, geologists have looked for a mechanism that would act slowly and gradually to part the continental land masses. Indeed, when Wegener proposed the concept in 1912, he wrote of 'wandering' continents, with its connotations of land masses floating gracefully about like icebergs on a calm ocean. But, as Cook, has pointed out:

> One can be sure that continents cannot really simply wander aimlessly over the surface of the Earth: exceedingly strong forces must be applied to cause them to move though the powerful ocean crust. In fact when they do move it is only under a force sufficient to fracture and plastically deform massive rocks of extremely high strength, a process that cannot occur uniformly but only in certain sudden, explosion-like processes.[5]

In developing a non-uniformitarian model, Cook has taken up the suggestion of Hapgood and Campbell[6] that thickening of the polar ice-caps puts stresses of the required magnitude on the land masses beneath. In Cook's model, Pangaea stretched from pole to pole. Build-up of ice at one or both poles finally snapped the crust and the corresponding pressure at the other pole helped determine the direction of the main fracture. The pressures on the crust were thus rather like those of on the shell of a hardboiled egg being squeezed at both ends in a vice. The fracture would have occurred rapidly and explosively and the subsequent movement apart of the newly formed continents would also have been rapid, rather than a slow drift.

There is evidence, says Cook, that the Wisconsin ice-cap (the ancient Arctic) suddenly disappeared roughly 10,000 years ago. This mass of ice amounted to some 100 million billion tons. For it to have been melted by the sun would have taken a minimum of 30,000 years even under ideal conditions, but in any case this would have been more than enough for the Earth to 'recover' naturally from being deformed since the 'relaxation time' of the Earth's crust is less than 10,000 years. In order to account for the persistence of the Wisconsin depression it is necessary to conclude that the ice was dissipated catastrophically. Independent evidence, according to Cook, confirms that the Earth's crust in the region began to rapidly uplift at the same time as the ice disappeared, some 10,000 years ago. The mass of ice and snow thus released formed the present Arctic and Atlantic oceans, whose water content agrees reasonably well with the mass of the Wisconsin ice-cap calculated from crustal depression data.

The ice-cap model does not have a direct bearing on the age of the Earth, since it can be argued that even if the continents did break up as recently as 10,000 years ago, the Earth might still be of very great antiquity – perhaps billions of years as Darwinists believe. But the model does have an important indirect bearing on methods of geochronometry because it challenges a key part of the Darwinian view of historical geology.

Darwinists believe that when the continents parted, more than 65 million years ago, the primitive mammals then in existence were separated geographically into distinct populations. In isolation, these populations are said to have evolved quite independently to become, on one hand, the marsupial mammals of Australasia, and on the other, the placental mammals of Europe and America. This process is said to have led to some remarkable similarities. An often cited example is the similarity of the Tasmanian marsupial wolf and the American and European timber wolf which are non-marsupial.

This parallel development or 'convergence' is seen by many Darwinists as important evidence in favour of the natural selection mechanism (and is an issue examined in detail in Chapter 14). Clearly, however, if only some 10,000 years (or even 100,000 years, or 1 million years) have elapsed since the continents have parted, then nothing like enough time has passed for any appreciable evolutionary change to have taken place by means of spontaneous genetic mutation, and Darwinists can no longer appeal to the separation of land masses to support their theory. They have also to account for the similarity of Australian marsupials and their placental counterparts elsewhere by some other mechanism.

So it is reasonable to say that the ice-cap model points to a 'recent' origin for life in the sense that it dramatically reduces the time scale in which certain key phases of evolution were formerly supposed to have occurred and rules out a mechanism that relies on chance mutation.

CHAPTER 7

Fashioned from Clay

The synthetic theory of evolution rests not on seven pillars of wisdom, but on a solitary monolithic support – the geological column that is displayed in textbooks, classrooms and natural history museums round the world.

As a teaching aid, as a powerful multi-layered symbol of world history, above all as a public relations tool for the general theory of evolution, the geological column has been a brilliant success. It has the answer to every question on evolution and the age of the Earth; it is the one thing every school child takes home from museum field trips.

The geological column is at one and the same time extremely simple and extremely complex. To begin with, the column was simply intended to represent the rocks of the Earth's crust, in sequence and roughly in scale with the depth of rock, using colours to represent rock types in much the same way that the London underground or New York subway are represented by coloured lines on the tube map.

Like the tube map, the geological column has taken on a meaning that is more literal than symbolic. The first change occurred when *relative* dates were added. The coloured layers ceased to represent rock formations and became historical periods.

Uniformitarians began to talk about the Cambrian 'period' or the Cretaceous 'period' instead of the Cambrian rocks or Cretaceous rocks. The next step was to assign *absolute* dates to various horizons within the column, for instance the primary rocks at the very bottom and the volcanic or igneous intrusions which appeared from time to time. And once the absolute dates were in place, and the sequence of rocks had become a chronology of the Earth, it was a natural final step to include family trees showing how the animal and plant kingdoms were related through common ancestors dominant in the various periods.

The confusing relationship between the substantive role of the geological column and its role as an evolutionary metaphor is so subtle that it often escapes notice entirely. The first and most important of these confusions occurred when historical order was assigned to the many coloured rock strata. For this simple act carried profound implications about *how* and *how quickly* those rocks were formed. If the Cretaceous 'period' lasted for 65 million years as evolutionists believe, then the chalk which was laid down during that period must have accumulated very slowly.

Some three-quarters of the Earth's land mass is covered by successive layers of sedimentary rocks – that is, rocks laid down under water and sometimes enclosing fossils. (The term 'rock' is used by geologists to denote not only hard substances like limestone or sandstone, but also clays, shales, gravels, sands and any other substantial deposit of waterborne material.) The conditions under which these deposits have been laid down are said to be analogous to the conditions which exist today; ranging from the ocean bottom, to the floor of shallow lakes; to coral reefs; to rivers and their estuaries or deltas. Conditions include both salt water and fresh water; tidal and non-tidal; inland sea and open ocean.

Some sedimentary rocks are called 'secondary' rocks because they are often composed of pieces of the primary or volcanic rocks that originally formed the Earth's crust. These pieces range from boulders and pebbles down to grains of sand and microscopic particles of silt and mud which at some time in the past were eroded from the crust and were transported usually by rivers into the oceans and later deposited to become new rocks.

The types of sedimentary rock these particles turn into depend on the kind of material from which they are made. Sand particles are compacted to become sandstones; pebbles are usually held together with finer material such as sand or silt to form a conglomerate. Silt and mud form fine-grained rocks such as shale or mudstone. Limestone, however, is not composed of particles of primary rock. Chalk, for instance, is a soft limestone said to be formed from a whitish mud or silt of organic origin.

While they were being deposited under water, these sediments often carried with them various kinds of debris, including the remains of animals and plants. The hard parts of these inclusions (such as bones, shells, teeth) often survive as fossils, sometimes being chemically altered to become stone-like.

The extent to which the remains were preserved by burial varies greatly, especially in the case of the soft fleshy parts which usually do not survive at all. In a few instances, though – such as the trilobites preserved in very fine-grained limestones – the detail 'recorded' in stone is almost miraculous and includes the microscopic crystalline structure of the eye. In rare cases, such as those of the famous Burgess shales of Canada, even soft-bodied animals are preserved as imprints in the rock.[1]

The various sedimentary rock strata are piled one on top of the other in chronological sequence, representing successive episodes or phases of deposition of sediment. These strata have been examined in great detail, extensively classified, and correlated with some precision all over the country and over the world. The chalk exposed in the sea cliffs of England, for instance, can be found across much of Northern Europe, from France as far north as Denmark. It is said to have been deposited at the bottom of a shallow sea – called the Cenomanian Sea from the Roman name for the French town of Le Mans – which covered much of Europe.

The study and interpretation of this sequence of sediments (the science of stratigraphy) is complicated by the fact that some of the beds have been laid down, only to be eroded again, giving rise to gaps in the sequence. As well, the Earth's crust has been much distorted by folding, and volcanic activity. What this means is that nowhere in the world is there known to be a complete sequence of sediments, from the oldest to the most recent, so that stratigraphy is largely a matter of comparison of one outcrop with another followed by inferences as to their relationship to each other and similar outcrops. Sometimes, this process is relatively simple, as in the case of the chalk cliffs of Kent which are cut through by the English Channel but which crop out in the same way, with the same fossils, on the coast of France. Sometimes, the comparisons are very much more difficult and may depend on complex and sensitive techniques like finding thin strata with a characteristic electrical resistivity or characteristic mineral content which can be traced from one country to another – or even one continent to another.

The most important technique used by geologists in studying the sequence of strata relies on the observation that many of the fossils contained in them appear to be restricted to one or a few particular sediments – or even a narrow band or 'horizon' within a sediment. For example, the chalk cliffs of Dover are characterised by certain species of fossil sea urchin, found in the chalk but not found elsewhere. Similarly the Oxford Clay – widely dug in the midlands of Britain to make bricks – is characterised by the fossilised shells of extinct shellfish related to the squid and known as ammonites.

Fossil species of this kind, which are believed to be associated uniquely with one type of sedimentary rock stratum, are known as zone index fossils and are used to identify that rock stratum whenever it is encountered. An example of the geological use of this technique is in drilling core samples from the sea bed when prospecting for oil or natural gas. If the core sample brought up contains remains of the sea urchin *Micraster* or the ammonite *Harpoceras* then the geologist knows he is unlikely to strike oil in that stratum, because they are zone index fossils from the chalk and the lias shales respectively – which rarely, if ever, contain oil.

Using these principles, uniformitarian geologists have constructed the geological column – a hypothetical sequence of all known sedimentary rocks from the earliest to the most recent, each sediment correlated with distinctive fossil remains that are deemed to illustrate the animal and plant life contemporary with each phase of sedimentation. The geological column is thus considered to show not only the 'record of the rocks' but also the 'fossil record' – that is to say the record of life on Earth from its beginnings to the present.

It is here that the use of the geological column as a metaphor for Darwinian evolutionary processes comes in. It is clearly of the greatest practical utility to be able to identify any given geological horizon by identifying the fossils it contains. Construction engineers building cuttings for roads and railways, petroleum geologists and many others employ this useful technique on a daily basis. It is of the greatest value to conjecture a 'perfect' chronological sequence of sediments. But once the animal remains associated with each deposit are piled one on top of each other, and once relative dates are assigned to those deposits, then the fossils in them cease to be dumb, inanimate

many more millions of years, the bituminous coal has been converted to anthracite: jet black, hard and burning fiercely when lit.

The geological events which gave rise to the formation of the largest coal deposits are said to have begun 325 million years ago and lasted for 45 million years, in a remote period named the Carboniferous, after its most characteristic product. The sediments laid down in this period are referred to as the coal measures and they are represented in many countries around the world.

The coal measures are immensely thick sedimentary deposits containing a variety of rock types, occurring in sequences which are often repeated. Typically, these sequences include beds of shale containing freshwater fossils; overlaid by strata of coal; overlaid in turn by thick beds of limestone containing fossils of marine animals. These repeated sequences, called cyclothems, are a key feature of the Carboniferous rocks and are always associated with coal deposits.

There are a number of important factors to notice about the uniformitarian scenario. A two-stage process is necessary to the theory in order to provide the initial high temperature to create the lowest rank of coal or lignin. It is necessary to the theory because the only other mechanism uniformitarians have available to accomplish coalification is a very slow rate of deposition of the overlying rock – typically 0.2 millimetres per year. Clearly this alone would not account for fossilisation of a forest. On the contrary, such low uniform rates of deposition would simply allow the trees to rot away and be dispersed by wave or current action.

A second necessary part of the theory is that the forested land must be inundated by the sea. This is necessary first because the sediments overlying coal strata contain marine fossils, but more significantly because uniformitarians need the pressure of accumulating overlying rocks to account for the slow transformation of peat into coal. For these reasons, uniformitarians suggest that coal-forming forests were on low-lying coastal plains or basins (usually in swampy conditions) which were subject to periodic marine invasion. This in turn means that there has to be periodic sinking or subsidence of the basin in which the forest grows, and this is said to be caused by major movements of the Earth's crust.

To summarise the uniformitarian coalification process: a forest grows up in a basin or plain beside the sea. The forest becomes swampy, but with fresh water. A vast peat bog forms. The Earth's crust shifts, the basin or plain sinks and the sea covers the peat bog. Over millions of years, limestone sediments are laid down on the bottom of the sea, compressing the peat bog and increasing the rank of coal thus formed. At the end of this period, the land rises and the basin or plain is exposed once more. Again, a forest springs up on the reclaimed land; the forest becomes swampy with fresh water; a vast peat bog forms. The Earth's crust shifts and the plain or basin sinks once again beneath the sea. More marine limestones are deposited, and so on.

If uniformitarians claimed that this had happened once, or even twice, or even three times, in the same spot, we would have to grant that their scenario could have occurred. But in the coal measures, this sequence is repeated not two or three times, but as much as *sixty* times.

According to Hollingsworth, writing on coal formation,

> In the case of the permo-Carboniferous [rocks] of India, the Barakar beds of the Damuda series, overlying the Tachir boulder bed, includes numerous coal seams, some up to 100 feet thick, occurring in a well-developed and oft-repeated cycle of sandstone, shale, coal . . . the vegetation is considered to be drift accumulation.
>
> The concept of periodic epirogeny is a reasonable one, but a more or less complete cessation of clastic [derived rock particle] sedimentation in the lacustrine basin during coal accumulation is difficult to account for on a wholly diastrophic origin. As an explanation for the fifty to sixty cycles of the Damuda system, it has an element of unreality.[1]

This 'element of unreality' also attaches to some other aspects of the uniformitarian view of coal origins. In 1945 Melvin Cook was appointed by the Unites States Navy to direct a high explosives group. Part of the group's work was to develop safer explosives for use in coal mining and as part of this project he made a special study of the occurrence and characteristics of coal. Cook points out that wood is composed mainly of cellulose with about one quarter a material called lignin. This chemical composition is the basis of wood (and coal's) usefulness as a heat source. Burning wood can give off almost as much heat on a weight-for-weight basis as a powerful explosive like TNT. According to Cook, the dehydration decomposition of wood gives rise to exothermal heat in the range 400 to 800 calories per gram, compared with TNT which generates about 1,000 calories per gram. If wood is compacted and put under pressure (as by being buried) the decomposition initiated by pressure alone will supply the necessary temperature to convert wood to coal, making the hypothetical biochemical or peat bog stage quite unnecessary.[2]

The best evidence that pressure, rather than time, is the cause of coalification comes from examining the rank of coal in relation to the depth of its deposit. In the United States, the Pittsburgh coal seam runs between Ohio and Pittsburgh and the strata in which the seam is contained dip downwards into the Earth at the rate of 20 to 40 feet per mile, with the coal at the easternmost end of the seam several thousand feet deeper than at the western end. As the seam goes deeper, the grade of coal increases: the deeper the burial and the greater the compression of overlying beds, the further the process of coalification has proceeded. In this case, the reaction would be started without any microbiological attack and could be achieved rapidly by pressure alone.

If coal were formed relatively quickly by rapid burial under marine sediments, then swamps, peat bogs, microbiological attack and millions of years of gradual deposition and slow pressure are no longer needed. Once again, as with the other sediments of the geological column, the key question is not so much how they were formed, but how quickly they were formed.

There is so much evidence on this point that it is hard to see how it can have been overlooked by uniformitarians. Although fossils are relatively rare in the coal itself, it is common for miners to discover large inclusions in the coal seams, such as boulders. And when fossils are found, they can be spectacular.

In 1959 Broadhurst and Magraw described a fossilised tree, in position of growth, from the coal measures at Blackrod, near Wigan in Lancashire:

> The tree was preserved in the form of a cast and the evidence suggested that the cast was at least 38 feet in height. The original

tree must have been surrounded and buried by sediment which was compacted before the bulk of the tree decomposed so that the cavity vacated by the trunk could be occupied by new sediment which formed the cast. This implies a rapid rate of sedimentation around the original tree.[3]

Broadhurst also says that such fossil trees in position of growth are far from rare in Lancashire and points out that Teichmuller in 1956 reached the same conclusions for similar trees in the Rhein-Westfalen coal measures of Germany.

In 1878 miners at Bernissart, a small village in the Mons coalfield of south-west Belgium, made a spectacular discovery when they uncovered a fissure in the coal seam packed full of complete dinosaur skeletons, at a depth of some 322 metres. Thirty-nine skeletons of the dinosaur iguanodon were recovered, from a fissure 100 feet high, many of them complete, and are today on display in the Royal Institute of Natural Sciences in Brussels.

The most striking thing about these creatures is that they measured 10 metres in length, stood several metres high and weighed in the region of two tons apiece. For their bodies to become rapidly buried would require rates of deposition thousands or even millions of times greater than the average 0.2 millimetres per year proposed by uniformitarians.

Geologists and palaeontologists are well aware from their field studies that the same thing applies to all of the sediments of the geological column. It is commonplace to find large fossils in position of growth or taking up their original volume, such as horsetails in Deltaic sandstones, corals in the Oolitic limestones, giant ammonites in the Portland beds and tree trunks in strata of many kinds.

Certainly the custodians of the geological column at London's British Museum of Natural History cannot fail to be aware of such discoveries. Each morning, on their way to work, they pass in the museum's grounds a fossilised tree trunk from the Lower Carboniferous, excavated at Craigleith quarry, Edinburgh, which originally measured some twenty feet or more in height.

CHAPTER 9

When Worlds Collide

As a journalist, I know only too well that you can't believe all you read in the newspapers. Yet I find one of our most interesting – and perhaps least-studied – psychological traits is not so much our gullibility as our propensity for ignoring facts which fail to fit with the conventional wisdom of our times.

One writer who challenged the received wisdom on geological sedimentation – and who was badly mauled for his trouble – was Immanuel Velikovsky, the American psychologist whose 1950 book *Worlds in Collision* caused something close to panic in the US academic community. Velikovsky proposed that a near collision between Earth and other planets of the solar system caused catastrophic geological events in the Earth's history and that most of the major features of the Earth's crust are testimony to these events. He also sought to establish a short timescale for the Earth's history.[1]

Velikovsky's book caused the sort of reaction amongst the American scientific fraternity that one might expect had he proposed the collision were actually going to happen next Friday. Geologists and astronomers were so virulently opposed to Velikovsky's book that they threatened to boycott the scientific textbooks of his publisher, Macmillan, forcing the firm to turn Velikovsky's work over to another publisher, not involved in textbook publishing (Doubleday).[2] Today, only forty years later, a concept closely similar to Velikovsky's is widely accepted by many geologists – that the major extinction at the end of the Cretaceous (and possibly other extinctions) was caused by collision with a giant meteor or even asteroid.

Velikovsky was so badly treated by the scientific community that he determined to back up his theory with an unchallengeable body of evidence. He spent some five years researching a further book on the catastrophist theme, *Earth in Upheaval*, in which he provides detailed evidence on scores of geological structures and palaeontological finds which are inexplicable on any basis other than a catastrophic origin. Moreover, the extent of the catastrophe required to produce these structures he showed to be global, and the energies needed on a cataclysmic scale.[3]

I will not be needlessly repetitive here of Velikovsky's very detailed research (his books are listed in the bibliography), but I will summarise three of his examples which are not only well attested from multiple sources, but which patently cannot be accounted for on any but a catastrophist model. They are the young age and rapid building of the world's mountain chains; the gigantic extent of certain rock formations, requiring singular, acute causes; and the occurrence of extinctions on a massive scale. There are also two mysteries which require explanation: anomalies relating to glaciation in the Ice Age, and the existence of beds of fossils thousands of feet deep but containing the remains of terrestrial, rather than marine, animals.

The major mountain chains are today believed to be the result of pressure at the edges of the continental 'plates': in effect they are the buckling of the edge of one plate by another – rather like two cars in a road accident. The Andes in South America and their North American counterpart, the Rockies, are said to be caused by pressure from the Pacific plate on the American plate. True to the uniformitarian model, this movement of plates and consequent mountain building is deemed to have taken place not at all like a traffic accident, but very slowly over millions of years at a rate in the order of 1 to 10 centimetres per year. The trouble with this idea is that there is a substantial body of evidence pointing to rapid mountain-building occurring in the recent past – thousands, rather than millions of years ago.

In the Alps, for example, there are numerous sites of human occupation at altitudes which must be far above their original level. Human artefacts dating from the Pleistocene or Ice Age have been discovered in caverns at Wildkirchli, near the top of Ebenalp, at 4,900 feet (nearly one mile) above sea level. Even more astonishing is the cavern of Drachenloch near the top of Drachenberg, south of Ragaz, which was also occupied by humans during the Pleistocene and is some 8,000 feet above sea level (well over a mile and a half high). There are other examples in other continents of routine human habitation at extraordinary heights, especially in the Andes. This appears to point decisively to a substantial part of mountain-building activity taking place in the recent past.

A study of the Ice Age in India by Helmut de Terra of the Carnegie Institution and Professor T. T. Paterson of Harvard University concluded that the Himalayas were still being built during the Ice Age, and reached their present great height only during the historical era. 'Tilting of terraces and lacustrine beds,' wrote de Terra in *Studies on the Ice Age in India and Associated Human Cultures* in 1939, indicates a 'continued uplift of the entire Himalayan tract' during the last phases of the Ice Age.[4]

At 12,500 feet up in the Andes (two and a half miles above sea level) is the deserted but well-preserved city of Tiahuanacu. It is in a region where corn will not ripen and its altitude is too high today to support life for anyone other than a tribe of mountaineers. The president of the Royal Geographical Society, Leonard Darwin, suggested in 1910 that the mountains had risen considerably after the city was built, and it is hard to find an alternative explanation that is credible. If the Andes were as little as three thousand feet lower, corn would ripen in the basin of Lake Titicaca and the site of Tiahuanacu would support a sizeable population.

The second indicator of catastrophism on a grand scale is the extent of certain geological formations, principally volcanic lava flows. In North America, an area of 200,000 square miles in Idaho, Washington State and Oregon, known as the Columbia Plateau, is covered by lava to a depth as great as 5,000 feet (almost 1 mile). Uniformitarianism could never account for such beds. This quantity of lava exceeds by many orders of magnitude all the lava flows from all the world's currently active volcanoes. And there are similar deposits on other continents, such as the Deccan traps in India, 250,000 miles square and several thousand feet deep, the lava bed of the Pacific Ocean and the lava dykes that cross South Africa.

The third indicator of historical catastrophes is that of extinctions on a huge scale. A common rock in the geological record is the Old Red Sandstone. The northern half of Scotland from Loch Ness to the Orkneys exposes this rock formation in myriad sites to a total depth of more than 8,000 feet (twice the height of

Ben Nevis). In an area 100 miles across, the Old Red Sandstone contains the fossils of billions of fish, contorted and contracted as though in convulsion and resulting apparently from some catastrophic event.

Describing the fossil fauna in his 1841 study, *The Old Red Sandstone*, Hugh Miller wrote, 'Some terrible catastrophe involved in sudden destruction the fish of an area at least a hundred miles from boundary to boundary, perhaps much more. The same platform in Orkney as at Cromarty is strewed thick with remains, which exhibit unequivocally the marks of violent death.'[5] The same picture is found at Monte Bolca in northern Italy where Buckland, writing in 1836, observed that 'the circumstances under which the fossil fishes are found at Monte Bolca seem to indicate that they perished suddenly. The skeletons of these fish lie parallel to the laminae of the strata of the calcareous slate; they are always entire and closely packed on one another . . . All these fishes must have died suddenly . . .'[6]

Similar formations are found in the coal measures of Saar-brucken on the Saar, the calcareous slate of Solenhofen, the blue slate of Glaris, the marlstone of Oensingen in Switzerland and of Aix-en-Provence. In the United States there are comparable formations such as the black limestones of Ohio and Michigan, the Green River bed of Arizona and the diatom beds of Lompoc, California. D. S. Jordan reported finding in the Monterey shale of California enormous numbers of the fossil herring *Xyne grex*. Jordan estimated that more than one billion fish averaging 6 to 8 inches in length died on four square miles of sea bed.[7] Ladd points out that catastrophic death of fish on a large scale sometimes does occur today, in the case of so called 'red water' for example. What does not occur, however, is death on a scale of billions. Nor do the victims become rapidly buried in thousands of feet of sediment and fossilised – their carcasses are preyed on by scavengers.

The two mysteries that have a bearing on catastrophes receive little publicity, yet are tantalising in the extreme, crying out for an answer. The first has to do with glaciation during the Ice Age. It was a Swiss naturalist, Louis Agassiz, who realised in the 1830s that much of Europe must once have been covered in ice. Agassiz became fascinated by glaciers in his native Switzerland. He even built a hut on a glacier at Aar and lived in it so he could study the movement of the ice front. He deduced from this study all the actions of which glaciers are capable: transporting large quantities of rocks and stones (including huge boulders) which will be left behind when the ice melts (moraines); gouging out U-shaped valleys; and cutting striations in the underlying rock surface.

Agassiz converted Dean Buckland, influential president of the Geological Society, to belief in an Ice Age by showing him distinctive glacial features in Scotland, and he converted Charles Lyell by showing him some moraines within two miles of his father's house. Having secured such powerful backing, Agassiz's theory was certain of universal acceptance. A key consequence of this widespread acceptance has been the tendency to ascribe *all* inexplicable terrestrial features to glacial action. This action, of course, is believed to have taken place according to the uniformitarian model over hundreds of thousands or even millions of years.

One particular feature which glaciation is used to explain is the occurrence of 'erratics' – substantial rocks which are geologically out of place. On the coast of Scotland are large quantities of rocks which have been transported from the mountains of Norway. In North America, erratic blocks of Canadian granite are found over ten of the northern United States from Maine to Ohio. All these are believed to have been moved by glaciers working slowly but surely.

And this is where the puzzles come in. In Eastern Europe, there are many erratics strewn over the Russian plains. But whereas in Finland and the northern provinces these blocks are large, they get uniformly smaller as one goes south. A similar pattern of uniform grading of erratics is found elsewhere in Europe and North America. This distribution points not to ice but to water action, and water on a huge scale. The uniform grading also points to turbulent flood conditions gradually abating.

In addition marine fossils are found *on top* of glacial deposits as in the case of the whale skeletons found in bogs covering glacial deposits in Michigan. Whale fossils have also been found 440 feet above sea level north of Lake Ontario, at Vermont more than 500 feet above sea level, and some 600 feet above sea level in the Montreal area, according to Dunbar in *Historical Geology*. As Velikovsky observes, 'Although whales occasionally enter the mouth of the St Lawrence river, they do not usually climb the surrounding hills.'[8]

The second mystery is one that has intrigued many geologists since the early nineteenth century, including Alfred Russell Wallace, co-discoverer with Darwin of evolution by natural selection. The mystery concerns a range of hills called the Siwalik Hills north of the Indian capital Delhi. The hills, some 2,000 to 3,000 feet high and several hundred miles long, are actually the foothills of the Himalayas. The Siwaliks contain extraordinarily rich beds crammed with fossils: hundreds of feet of sediment, packed with the jumbled bones of scores of extinct species. Many of the creatures were remarkable, including a tortoise twenty feet long and a species of elephant with tusks fourteen feet long and three feet in circumference. Other animals commonly found include pigs, rhinoceroses, apes and oxen. Most of the species whose fossils are found are today extinct, including some thirty species of elephant of which only one has survived in India.

Beds of this sort are common in the geological record, as in the case of the fish beds referred to above. But the Siwalik beds contain the remains of *terrestrial* animals, not marine creatures. These animals must have been killed by some singular event over a relatively short space of time, and an event which took place on land. And whatever the nature of the event, it resulted not only in catastrophic extinction of many species but also the formation of beds of sediment thousands of feet thick.

It is sometimes suggested that the animals were killed by the onset of the Ice Age, but no mechanism has been proposed that would account for such large numbers being killed by ice creeping along at a few centimetres a year, or for their rapid burial in hundreds (sometimes thousands) of feet of sediment.

It was also proposed that the Siwalik deposits were alluvial, and represented debris carried down by the torrential Himalayan streams. But it was realised, as Wadia wrote in his *Geology of India*, the alluvial explanation 'does not appear to be tenable on the ground of the remarkable homogeneity that the deposits possess' and their 'uniformity of lithologic composition' in many different and isolated basins miles apart.[9]

Thirteen hundred miles from the Siwalik hills in central Burma are deposits of a very similar nature, containing remains of mastodon, hippopotamus and ox along with large quantities of fossil wood – thousands of fossilised tree trunks and logs scattered in the sandstone sediments. In total the deposits may be as much as 10,000 feet thick and there are two distinct fossiliferous horizons separated by 4,000 feet of sandstone.

Graveyards of terrestrial animals are commonplace. In his books on dinosaurs, Dr Edwin Colbert gives numerous examples. In New Mexico 'There were literally scores of skeletons on top of one another and interlaced with one another. It would appear that some local catastrophe had overtaken these dinosaurs, so that they all died together and were buried together.'[10]

At Como Bluffs, Wyoming, referred to earlier, 'The fossil hunters found a hillside literally covered with large fragments of dinosaur bones . . . In short it was a veritable mine of dinosaur bones.'

In Alberta, Canada, 'Innumerable bones and many fine skeletons of dinosaurs and other associated reptiles have been quarried from these badlands, particularly in the 15–mile stretch of river to the east of Steveville, a stretch that is a veritable dinosaurian graveyard.'[11]

Colbert refers also to the Belgian dinosaur find mentioned in the last chapter: 'Thus it could be seen that the fossil boneyard was evidently one of gigantic proportions, especially notable because of its vertical extension through more than a hundred feet of rock.'

These examples can be multiplied almost endlessly, yet few modern geologists are prepared to accept that major features of the Earth's crust could have been caused by singular events, because such an admission seems to open the door to some kind of geological anarchy which threatens the orderly arrangement of exhibits in their glass cases.

The first half of this book has been concerned with the purely geological question of how, and how fast, the rocks of the geological column were formed. An even more significant question from the point of view of synthetic evolution theory is what conclusions about the origin of life can be drawn from the fossils contained in the geological column. To seek answers to this question we must turn from geology to its close relative, palaeontology.

PART THREE

Chance

CHAPTER 10

The Record of the Rocks

One afternoon in 1822, a medical practitioner from Brighton, Dr Gideon Mantell, took his wife for a walk in the Sussex beauty spot of Ashdown Forest. Mrs Mantell must have been unusually observant for she picked up a strange tooth from a pile of road-mending stone they passed. Puzzled by the tooth, Mantell showed it to the leading geologist of his day, Charles Lyell (who in turn showed it to France's Baron Cuvier), but neither was able to identify the animal from which it had come. Mantell turned to his own professional body, the Royal College of Surgeons, and made a systematic search through the College's collection of teeth but without success. He was about to give up when the curator showed him a newly discovered lizard specimen just arrived from America, called an iguana. By an extraordinary coincidence, its teeth were closely similar to that found by his wife, and Mantell realised he was holding the tooth of an unknown extinct reptile of great size. The tooth was described by William Conybeare, who coined the name iguanodon to describe its long dead owner – the first dinosaur to be identified.

The tale of this seminal event is in some ways a parable for the history of palaeontology as a whole. Comparative anatomy has played a decisive role in the development of the science; much can be deduced from apparently meagre finds, as long as their significance is appreciated; chance plays a substantial part in palaeontological discoveries; and a little intelligent guesswork can often go a long way to solving a tantalising mystery.

But there is a negative aspect to the parable. His discovery made Gideon Mantell obsessed with finding more dinosaur remains and he filled their Brighton home with so many rock specimens that his wife left him, never to return. This obsession with fossils is by no means rare and has something in common with the Gold Fever experienced by prospectors. It is an obsession that has played a part in palaeontology and evolution theory in many ways over the past 150 years.

The modern, stratigraphical, significance of fossils came about through the large-scale engineering works undertaken at the beginning of the nineteenth century. Hundreds of miles of roads, canals and railway cuttings were dug across Britain exposing rocks of every kind along with the fossils they contained. William Smith, the 'Father of English Geology', was an engineer responsible for cutting canals and he noticed that similar sequences of rock strata recurred in different places and that they often contained similar fossils. Wherever his navvies' picks struck the creamy rocks of the Lower Oolites, there he found the distinctive mollusc *Trigonia*. When they dug the tough blue shales of the Lias, there he found the oyster *Gryphaea* called 'devil's toenails' by country people.

Smith began to draw up the first geological maps – of the countryside around the city of Bath – marking the different beds of rock in different colours. Such charts were immensely useful to him in siting and digging his canals: telling where to find building stone for aqueducts and bridges; where to find clay to waterproof his canal across porous rocks; which line to take across the countryside. It was largely because of his unaided pioneering efforts in publishing the first geological map of Britain in 1815 that the Geological Survey of Great Britain was established soon afterwards.

Smith made one further observation about the fossils he discovered. He realised they were the remains of marine creatures and that the rocks he was looking at were the floor of an ancient sea – in fact a succession of such seas – which had covered Britain some time in the past. He realised, too, that the succession of rocks was accompanied by a succession of fossils – the earliest at the bottom, the newest at the top.

Smith's legacy to geology is of incalculably great value. Unfortunately, when he decided to distinguish different rock formations by different colours in his first maps, he unwittingly bequeathed geology a technique that was to have unexpectedly ambiguous results later on. For he made it possible – indeed, almost inescapably natural – to associate a *chronological succession* of rocks with an *evolutionary succession* of life forms in the past.

Smith had been interested in fossils since he was a young boy living on a farm near Oxford and had amassed a large collection of fossil specimens. As a result of his researches as a canal engineer, he rearranged his collection of specimens stratigraphically: all the fossils from the Bradford clay in one drawer, all those from the lias shales in another, and so on. It might be thought that this is an obvious thing to do, but in fact it is more natural to arrange fossils collected from different locations in *biological* groups, for comparison – all the shark's teeth together, all the sea urchins together, and so on.

When such a biological arrangement is made, a great many species are found to be very stable in shape and size. An oyster from the Jurassic period looks very much like an oyster from any later period, including one washed down with champagne today in the Savoy Grill. On the other hand, one of the main reasons for supposing, as Darwinists and other evolutionists do, that species evolve as you proceed up the geological column is that some types of creature disappear and are replaced by something similar yet distinctively different. For instance, if you walk along the beach from Lyme Regis in Dorset to the neighbouring town of Charmouth, you will find that the rocks in the cliffs have been tilted at an angle by earth movements. As you travel along the beach you are able to look higher and higher up the geological succession. What you find in those rocks are different species of ammonite (a spiral shellfish related to the present day pearly nautilus) as you pass along. In the lowest bed is a genus called *Asteroceras*; in the next, one called *Amaltheus*; and in the highest *Harpoceras*.

Uniformitarian geologists believe these rocks took millions of years to form, and Darwinists say that the successive ammonite species represent a line of descent: that the ammonite *Harpoceras* near present-day Charmouth is the remote offspring of *Asteroceras* at the Lyme Regis end. The fact that there are gaps in the fossils and no transitional forms intermediate between the various species does not alter this conviction. Because the rocks are a succession and took millions of years to lay down, then the fossils they contain are a living succession also.

In one sense, it is very surprising that uniformitarian

geologists should think this way about the biological past, since it is quite antithetical to that most fundamental principle of geological history: that the past can be understood in terms of the present. For in the animal world, the most striking thing about species today is their discontinuity. The living world consists mostly of gaps between species; gaps that remain unbridgeable even in imagination. The fossil record indicates clearly that the living world also consisted of gaps in every past age from the most recent to the most remote. Yet Darwinists believe that while the present consists of gaps, the past was a perfect continuity of evolving species – even though this continuity is not recorded in the rocks – and they have bent immense efforts to establish this with credible sequences of fossil ancestors and descendants.

Probably the best known of such sequences is that of early horses discovered mainly in North America. Illustrations of this sequence figure prominently in textbooks on palaeontology and in natural history museums around the world. It is due to the passionate bone collecting of O. C. Marsh, Professor of Palaeontology at Yale University, and his intense rival Edward Cope. Their materials were arranged by Henry Fairfield Osborn, Director of the American Museum of Natural History, and his student William Matthew. As early as 1874, Marsh declared that 'The line of descent appears to have been direct and the remains now known supply every important form.'[1]

The sequence begins with a tiny creature the size of a fox terrier, poetically named *Eohippus* (Dawn Horse) by Marsh, with four toes on its front legs and three toes on its hind legs, and teeth suited to forest browsing. This creature is said by Darwinists to come from the lower Eocene, about 50 million years ago by their dating methods. In beds of Oligocene age (around 30 million years old) are found remains of *Mesohippus*, the size of a sheep, with three toes on each leg. In Miocene beds said to be 15 million years old are found fossils of *Merychippus* whose teeth are adapted to grass-feeding habits, still with three toes but walking on tiptoe. And from Pliocene beds of around 7 million years ago come remains of *Dinohippus*, an animal the size of a small pony with only one toe (hoof) with rudiments of toes either side, and teeth fully adapted to grazing.[2]

There is no question that these remains, when placed together, are strongly suggestive of an evolutionary development. They show what the evolutionary model predicts the fossil record should show. In fact, it is comparative anatomy of this sort that provides the strongest body of evidence for evolution in the first place, and it is easy to understand the enthusiasm and speed with which the American Museum set up its display – a display rapidly copied by the British and other museums.

From a purely scientific standpoint, however, there are two difficulties with this sequence. The first is that although the fossil record has been bountiful enough to provide these intermittent remains, it has been consistently reluctant to yield up any remains that are actually transitional between them. The similarities between *Eohippus* and *Mesohippus* are great. But their differences are greater still. Bones of *Eohippus* and bones of *Mesohippus* have been found in a number of places. But bones of the animals that are said to connect them in lineal descent are not merely rare – they are non-existent. And the same thing is true for most of the animals in the sequence: that transitional species are not merely unusual; they are missing entirely.

The second difficulty is that, given the continued existence of gaps in the fossil record, and the continued failure to find fossils of the hypothetical intermediate species, then to call the *Eohippus* sequence an evolutionary series is not a scientific theory – it is an act of faith, a matter of belief. It is perfectly true that an intelligent rational person can examine the remains and be convinced that they represent an evolutionary sequence, but not by virtue of any evidence that has been adduced, since the *Eohippus* sequence is not evidence for evolution. It is evidence for the former existence of different species of quadruped with a striking similarity, not evidence of a relationship between them. And it is this, the relationship – if any – which is the very matter in question.

According to Professor Garret Hardin,

There was a time when the existing fossils of the horses seemed to indicate a straight-line evolution from small to large, from dog-like to horse-like, from animals with simple grinding teeth to animals with the complicated cusps of the modern horse. It looked straight-line – like the links of a chain. But not for long. As more fossils were uncovered, the chain splayed out into the usual phylogenetic net, and it was all too apparent that evolution had not been in a straight line at all, but that (to consider size only) horses had now grown taller, now shorter with the passage of time. Unfortunately, before the picture was completely clear, an exhibit of horses as an example of orthogenesis had been set up at the American Museum of Natural History, photographed, and much reproduced in elementary textbooks (where it is still being reproduced today).[3]

Hardin was writing in 1961, but regrettably in view of what we now know, the same basic display is still on view at the British Museum of Natural History, making basically the same claims, and is still being reproduced in textbooks, and the current edition of *Encyclopaedia Britannica*.

One of the principal modern champions of Osborn's evolutionary sequence for horses has been George Simpson, who himself made important fossil horse discoveries in Texas in 1924, and whose 1951 book *Horses* first encapsulated all the findings of the American Museum team. The book makes fascinating reading, yet its author seems unaware of the many contradictions it contains. On the general question of horse evolution, Simpson says, 'The history of the horse family is still one of the clearest and most convincing for showing that organisms really have evolved . . .' and, 'There really is no point nowadays in continuing to collect and to study fossils simply to determine whether or not evolution is a fact. The question has been decisively answered in the affirmative.'

Compare this certitude with the following selection of quotations from the same book by Simpson (the paragraphs are quoted in sequence but are not connected in the original):

> Bonediggers have not, as yet, had the good fortune to find the precise immediate ancestors of eohippus or those that would show exactly where and when the horse family first arose.

> In Europe there are no really good collecting fields of early Eocene age and fossils are few, but eohippus forms a considerable percentage of those that are known. In the richer early Eocene beds of North America . . . eohippus is an abundant fossil. Hundreds . . . of specimens have been found, although most of them are fragmentary, single teeth or scraps of jaws or other bones. For some reason not clear to me, common as eohippus remains are, it is most unusual to find so much as a

whole skull and skeletons anywhere near complete are exceedingly rare. As far as I know, only four skeletons have ever been reconstructed and mounted.

It happens that fossil mammals from around the very end of the Eocene and the very beginning of the Oligocene have not been well known in America. In recent years this gap in knowledge is being filled, but we still do not know enough about the animals of that important time of transition from one epoch to another. This applies also to the horses, and around this time there is a slight break in our otherwise practically continuous knowledge of horse history.

The teeth of this horse [*Epihippus*] were more progressive than those of any typically Eocene form and more primitive than any of unquestioned Oligocene age, but somewhat nearer the earlier type. The skeleton is practically unknown and we can only guess that, when discovered, it may more fully confirm the reasonable inference that American Oligocene horses were directly derived from *Epihippus*. It remains possible however, that the immediate ancestor of the Oligocene horses lived in some other region where its bones have not been found.

One other peculiar and extinct group should be mentioned . . . the pygmy horses of the Miocene. These are united under the name *Archaeohippus* . . . It is a pity that the skeleton is so incompletely known that no mounted specimen or restoration can yet be made. At any rate this reversal of the usual, but by no means constant, tendency for horses to increase in size is of extraordinary interest.

Perfectly preserved feet of *Pliohippus* are rare and the tiny bones of the reduced side toes are readily lost.

Simpson concludes his book with what seems to me a remarkable statement of the rarity of all the finds on which the lineage of horses is based. Under the heading 'Where to see fossil horses' he writes:

Complete mounted skeletons of fossil horses are rarities. They are seldom found and their preparation is a long, laborious, highly skilled and expensive job. Much the greater part of the display and research materials of fossil horses consists of partial skeletons or, especially, isolated bones, skulls, jaws and lesser fragments. Of these partial fragments tens of thousands are known. Of mounted skeletons, there are fifty-odd in the United States . . . As far as I know, there are no mounted skeletons of *Epihippus, Archaeohippus, Megahippus, Stylohipparion, Nannippus, Calippus, Onohippidium,* or *Parahipparian,* and none in the United States of *Anchitherium* or *Hipparion.*[4]

Of course, it is not essential to have a complete skeleton in order to describe an extinct creature anatomically with a high level of confidence. It is, however, more than a little disturbing to learn that the descent of horses is being offered as *the* decisive evidence in favour of evolution on the strength of so little real physical evidence, and with so many gaps filled only with speculation. It is especially worrying, for instance, to learn that there are no fossils known of the creature that is said to have preceded *Eohippus* and that there is also a gap in the proposed sequence immediately after *Eohippus* and before its proposed descendant *Miohippus.* We are entitled to ask: what exactly is it that connects them scientifically?

The problems that have bedevilled horse palaeontology also beset every other branch of the science. Indeed, the gaps in the fossil record are reflected in the living world where many major animal and plant groups are high and dry with no discernible predecessors. The development of the entire order of mammals is missing from the fossil record, from its supposed shrew-like ancestor of the late cretaceous until modern times.

Palaeontologists have found one spectacular fossil which is said to be a clear example of transitional form not merely from one species to another, but from reptiles to birds: the famous *Archaeopteryx* skeletons found in the slates at Solenhofen in Bavaria. So rare and precious are the two chief specimens said to be that they are kept under guard in a bank vault, safe from harm at the hands of outraged creationists who, presumably, are thought to be plotting the theft or destruction of what they consider to be Darwinist forgeries!

Like the fossil horses, *Archaeopteryx* is an important discovery and one that appears to confirm the predictions of the Darwinist model. It seems to offer substantial evidence of a transitional form and together with fossil horses forms the centrepiece of most museum displays and textbook accounts. Like the horses, however, *Archaeopteryx* has substantial problems, and these have been compounded by more recent discoveries.

Darwinists believe *Archaeopteryx* is proof of a number of important parts of their theory. First it is said to demonstrate the existence of a feathered creature long before the age of birds – *Archaeopteryx* dates from the age of reptiles. Next it is said to have vestigial characteristics from its reptilian ancestors: claws on its feathered forelimbs, teeth in its beak, and a bony reptile-like tail. It indisputably possesses true feathers and wings, but it does not possess the large pectoral muscles and deeply keeled breast bone that would enable it to fly. It must thus have been either virtually flightless like a chicken, or have been a glider – perhaps a precursor of true flight, say Darwinists.

Like the horse fossils, all this seems very convincing – until you subject the claims to a detailed examination. The idea that *Archaeopteryx* had descended from dinosaurs was first floated by Darwin's champion Thomas Huxley in the 1870s because of the persuasive similarities of the legs and hips of birds with those of dinosaurs. However, in offering *Archaeopteryx* as a descendant of dinosaurs, Huxley was ignoring one important inconvenient fact – the fact that *Archaeopteryx*, like all birds, has a wish-bone (analogous to the mammalian clavicle or collar-bone) whereas dinosaurs did not possess collar-bones.

In 1926, Darwinist palaeontologist Gerhard Heilman published a very detailed review of all the evidence for bird origins and carefully analysed all the relevant anatomical questions. Heilman concluded that the most likely candidates for the ancestor of *Archaeopteryx* were dinosaurs and that among these, the 'coelosaurs' (small bipedal carnivores) were the best candidates. Unfortunately, Heilman wrote, dinosaurs did not have collar-bones so a coelosaur ancestry was out of the question. Heilman proposed, therefore, that *Archaeopteryx* must have descended from a hypothetical pre-dinosaur ancestor from the Triassic period, and later developed fused collar-bones by 'convergent' evolution. This conclusion – despite its wholly hypothetical foundations and somewhat circuitous logic – passed into Darwinist lore and has been repeated in textbooks and museum displays ever since.

Matters rested there until 1973, when Professor John Ostrom resurrected the idea that birds have descended from coelosaurs. Ostrom made a detailed anatomical analysis of *Archaeopteryx* and found some twenty points of similarity with coelosaurs. Moreover, further collecting had shown that a few dinosaurs did exist with collar-bones, so perhaps some coelosaurs or close relatives might have had collar-bones too. However, according to Dr David Norman in his *Encyclopaedia of Dinosaurs*,

> Dr Sam Tarsitano and Dr Max Hecht are recent advocates of Heilman's original proposals of a more distant Triassic archosaur ancestor of birds. They claim to have found major faults with Ostrom's original work. Also several embryologists claim that the three fingers of the modified hand of living birds could not possibly have evolved from the three fingers of the theropod hand because the hand of birds is composed of the 2nd, 3rd and 4th fingers while in theropods the fingers are the 1st, 2nd and 3rd! Quite where this leaves *Archaeopteryx*, which also appears to have a theropod like hand of fingers 1, 2 and 3, is a matter of some embarrassment – does it mean that *Archaeopteryx* was merely a feathered dinosaur and not related to birds at all?

Archaeopteryx has recently been subjected to even more embarrassment since it has been knocked off its perch as the earliest bird (if bird it is). Sankar Chatterjee, Professor of Palaeontology at Texas Tech University, described a newly discovered fossil bird in the July 1991 *Philosophical Transactions of the Royal Society*. The new fossils, called *Protoavis texensis*, are those of a creature the size of a pheasant which was undoubtedly capable of flapping flight. They come from beds in Texas said to be 75 million years older than those in which *Archaeopteryx* was found.[5]

This means that true birds, essentially the same as modern birds, were flying happily in the skies of Texas during the period that Darwinists like to call the age of reptiles; a further indication that their geochronometry may well be faulty and that birds and extinct reptiles were in fact contemporary in a more recent past.

There are other problems with *Archaeopteryx*. The possession of claws on its wings is not diagnostic of reptilian ancestry, nor is it unique to *Archaeopteryx* since there is a modern bird in Venezuela, the hoatzin, which has such claws on its winged forelimbs when young. The wing claws of both *Archaeopteryx* and the hoatzin are referred to by Darwinists sometimes as 'vestigial' but no evidence has been produced of what creatures they are descended from and hence what precisely the claws are vestiges of. Teeth and a bony reptile-like tail certainly are unique characteristics for a bird, but it is no longer certain that *Archaeopteryx* is a bird.

So although *Archaeopteryx* is undoubtedly a fossil discovery of some significance, it is quite impossible to say at present exactly what that significance is. More importantly, it is impossible for Darwinists to claim that it supports the mechanism of random genetic mutation coupled with natural selection. *Archaeopteryx* provides no evidence at all for either mechanism, since it is completely isolated in the fossil record, just like *Eohippus*, with no known direct predecessor and no known direct descendant.

Synthetic evolutionists have dealt with the lack of real transitions in the fossil record in two ways, both of which are perfectly reasonable. First, they have said that all vertebrate fossil remains are relatively rare and finds depend largely on chance. The fact that a particular specimen has not yet been discovered does not rule out the possibility of its being found at some future date.

Darwin himself raised this point in connection with the lack of fossil remains of early humans, showing part-ape part-human characteristics, when he observed in *The Descent of Man* that,

> With respect to the absence of fossil remains serving to connect man with his ape-like progenitors, no-one will lay much stress on this fact who reads Sir C. Lyell's discussion, where he shows that in all the vertebrate classes the discovery of fossil remains has been a very slow and fortuitous process. Nor should it be forgotten that those regions which are the most likely to afford remains connecting man with some extinct ape-like creature, have not as yet been searched by geologists.[6]

In fact, more than 100 years of further intense collecting by well-funded professional expeditions has not yet yielded any of the remains that Darwin envisaged and Africa and the Middle East (the areas 'most likely') have now been thoroughly searched. There are early ape-like remains and there are early hominid remains. Indeed the store of primate fossils has been multiplied a thousandfold since Darwin. But the only 'missing link' so far discovered remains the bogus Piltdown Man, where a practical joker associated the jaw of an orang-utan with the skull of a human.

Darwin also gloomily confessed in *The Origin of Species* that,

> The number of intermediate varieties which have formerly existed on Earth must be truly enormous. Why then is not every geological formation and every stratum full of such intermediate links? Geology assuredly does not reveal any such finely gra- duated organic chain; and this, perhaps, is the most obvious and gravest objection which can be urged against my theory.[7]

Some evolutionists have explained the absence of transitional remains by suggesting that evolution proceeds in fits and starts. A species like *Eohippus* could remain stable for a long time – perhaps millions of years – thus giving rise to many individuals, some of whose bodies are fossilised. But then there is a spurt of evolutionary activity and *Eohippus* relatively quickly mutates into *Mesohippus* which again remains stable for millions of years and gives rise to many fossil remains.

Again, any reasonable person can hold this view quite prop- erly. But as before, he cannot hold it by virtue of the evidence of the geological record, because there is no palaeontological evi- dence for such evolution in bursts – except the lack of transition- al fossils which was the very reason for the existence of this point of view. The solution begs the question.

An exasperated Darwinist may well feel entitled to ask: if you won't accept the *Eohippus* sequence or *Archaeopteryx* as evidence for transitions, what on earth will you accept?

The answer is extraordinarily simple. Three-quarters of the Earth's surface is covered with sedimentary rocks. A great proportion of these rocks are continuously stratified where they outcrop and the strata contain distinctive fossils such as sea urchins in the chalk and ammonites in most Mesozoic rocks. The case for Darwinism would be made convincingly if someone were to produce a sequence of fossils from a sequence of adjacent strata (such as ammonite species or sea urchins) showing indisputable signs of progressive change on the same basic stock, but above the species level (as opposed to subspeci- fic variation). Even better would be a long sequence – eight or ten or twenty successive fossils – showing major generic evolu- tion, but a short sequence would be enough.

But this simple relationship is not what is shown in the sequence of the rocks. Nowhere in the world has anyone met this simple evidential criterion with a straightforward fossil sequence from successive strata. Yet there are so many billions of fossils available from so many thousands of strata that the failure to meet this modest demand is inexplicable if evolution has taken place in the way Darwin and his followers have envisaged. It ought to be relatively easy to assemble not merely a handful but hundreds of species arranged in lineal descent. Primary school children should be able to do this on an afternoon's nature study trip to the local quarry: but even the world's foremost palaeontologists have failed to do so with the whole Earth to choose from and the resources of the world's greatest universities at their disposal.

In Kent there is quarried a stiff blue clay called the gault which has been dug by brickmakers for hundreds of years and which, once fired, gives London's Georgian houses their distinctive yellow brick. This useful deposit is also important historically.

The authors of the *Geological Survey of Great Britain* memoir on Folkestone say:

> Nowhere is the gault more readily accessible, its fossils more abundant or more perfectly preserved than in the cliffs and shore of Folkestone. The descriptions of the gault cliffs at Copt Point by De Rance and Price constituted one of the earliest uses of palaeontology for stratal subdivision on modern lines. Added to the more recent researches of L. F. Spath, these investigations have raised the gault succession at Folkestone to the status of an international yard-stick for middle and upper Albian times.[8]

The authors then reproduce a detailed table of ammonite zones compiled by Spath (1923 and 1942) and modified by Breistroffer (1947) and Casey (1949). This table lists fourteen successive beds distinguished by ammonites in four major zones. The four zones are called after the ammonites they contain: the lowest zone is called the *dentatus* zone, the next the *lautus* zone, the third the *inflatum* zone, and last the *dispar* zone. The species of ammonite associated with these zones can be collected by the thousand. Museums and private collections are full of them, preserved in beautiful detail including an iridescent pearly shell. They come from a section of clay perhaps 100 feet high, which presumably, in uniformitarian terms, represents millions of years of sedimentation. Yet among the thousands and thousands of specimens dug up by collectors, no one has ever found a specimen that is part way between *Hoplites dentatus* and *Euhoplites lautus* or between *lautus* and *Mortoniceras inflatum* – or between any of the fourteen different ammonites.

There are plenty of other ammonites in the clay, of all shapes and sizes – often described by Darwinist geologists as separate species. Unfortunately, however, these do not fall neatly in between the primary species in their anatomy, nor neatly in the clay strata between them. They do not show evolution in a straight line but, like the horses, 'fall into the usual phylogenetic net'. Ammonites get more ribs then less ribs; they become closely coiled then loosely coiled; they grow lumps then they become smooth, then they grow lumps again.

This one example can be multiplied by all the quarries and all the sea cliffs, all the road cuttings and canal ways and railway embankments in the world. Wherever there are successive strata containing distinctive species, no one has ever demonstrated an unmistakable line of descent. Indeed, one thing that becomes plain to the open-minded geologist as he travels from exposure to exposure in search of fossils is that nature almost perversely precedes and follows one species by quite different ones.

Some Darwinists have even attempted to press this perversity into acting as evidence for their theory. Consider this example. In the Cotswold hills, near Gloucester, is a large brickpit at the village of Blockley. The bluish clay at Blockley looks rather like the gault of Folkestone but is actually an earlier formation known as the lias, dating from the Jurassic period. The liassic clays at Blockley provide very many well preserved ammonites which are used as zone index fossils.

There are two main kinds of ammonite found at Blockley. There are fat ones with two rows of knobs on the side (called *Liparoceras*) and thinner ones with no knobs (called *Aegoceras*). Occasionally collectors have also found a third kind which is said to be intermediate between these two and which has been called *Androgynoceras*. This third kind resembles *Aegoceras* in its inner whorls (that is, when it was young and the shell was first forming) but later on resembles *Liparoceras* with its two rows of knobs.

In 1870 a dedicated Darwinist called Hyatt took the specimens in the British Museum and arranged them in the order *Aegoceras* (oldest)–*Androgynoceras*–*Liparoceras* (youngest). His reasoning was based on the then fashionable evolutionary theory of 'recapitulation', the supposed repetition of characteristics in later generations. In 1938, the distinguished palaeontologist L. F. Spath took the same specimens from the British Museum and after careful examination reversed the order of evolution: *Liparoceras*–*Androgynoceras*–*Aegoceras*. Careful collecting in Dorset had revealed that the oldest (lowest) beds yielded only *Liparoceras*, while *Aegoceras* was found only in the youngest beds, accompanied by occasional *Liparoceras*. In reaching this conclusion, Spath was employing another evolutionary concept that was very different from that used by Hyatt. This time, it was called 'proterogenesis' and was said to concern the appearance of new characteristics only in the young, which then later spread to the outer parts of the shell.

In 1963 the ammonites from Blockley, where all three types are found in sequence, were examined again by Callomon. On this occasion, neither of the previous explanations was found to be satisfactory and the variation of *Androgynoceras* was attributed instead to 'extreme sexual dimorphism' – meaning that male and female of the same species were of radically different shape.[9]

One sympathises, naturally, with the difficulties of the palaeontologist attempting to come to grips with the sex of a creature whose last night on the town (according to uniformitarians) was some 150 million years ago. No one can blame a researcher who makes a mistake that is rectified by further research, for this is the very method of science. What I believe the Blockley ammonites demonstrate is something else. It is a prime illustration of the infinite elasticity of Darwinian theory: of its ability to interpret the data in any one of a number of completely different ways – even with diametrically opposed conclusions – as long as those ways are consistent with the central belief in Darwinian evolution itself.

'Recapitulation' means the fossils evolved one way. 'Proterogenesis' means they evolved in the opposite direction. In reality, the Blockley ammonites give no clue to lineage at all, just like all the other ammonites from all the other quarries.

Probably the most ambitious and comprehensive work on palaeontology ever to be published is the series of volumes produced in the 1950s by the Geological Society of America and the University of Kansas Press under the guidance of a committee of the most distinguished palaeontologists in the English-speaking world. Under the title *Treatise on Invertebrate Paleontology* some twenty-four volumes draw together the sum of human knowledge on thousands of fossil species. If solid fossil evidence of evolution is to be found at all, it is to be found documented here, in the volume dealing with the richest of all fossil fauna, the ammonites. In Volume L are illustrated and described in minute detail hundreds of ammonite species. Yet under the promising heading of 'Iterative Evolution', we find the authors glumly admitting:

> The complexity of the modern classification, seen in the systematic parts of this *Treatise*, results mainly from acceptance in large measure of the theory of iterative evolution, although there are relatively few proved examples of its occurrence. We often feel sure that it has occurred and that we should be deceived if we accepted similarities at their face value (Haas, 1942 has brought together a few outstanding examples) but we can seldom demonstrate just what the 'iterative' relationships are.

A little later, under the heading 'Examples of ammonoid evolution' the editor issues this warning to readers keen to learn what proof the fossil record has to offer:

> Waagen (1869) in a pioneer work attempted to demonstrate lineages, or lines of descent . . . The chief obstacle to such studies is that a lineage is an oversimplified concept; it is impossible to pick out a stratified succession of individuals which can with certainty be said to be genetically connected in the strict ancestor-descendant relationship.[10]

Having warned readers that the process is 'impossible', the editor then moves on to quote the works of Spath and Howarth as making just such a connection. As is always the case, the descriptions, the evolutionary 'reasoning' offered and the suggested lines of descent are unconvincing.

Of course, this is not to say that the findings of comparative anatomy are without foundation – quite the contrary. It was the very fact that animals as diverse as the mouse and the elephant both display a similar four-limbed anatomical pattern that first led biologists to think that they might have a common ancestor. As well as the obvious physical similarities between so many diverse creatures suggesting family relationships, there is the fact that very similar anatomical features seem to have been arrived at quite independently in different creatures adapted to similar modes of life (a phenomenon termed 'convergence' by Darwinists). Wings, for example, appear to have evolved quite independently no less than four times: in insects, in flying reptiles; in birds; and in bats.

These and many other examples of similarities in form seem to point unequivocally to common ancestry and common processes of evolutionary change – until they are re-examined in the light of the scientific paradigm in which they are tacitly contained; the ruling ideology that provides the common theme for the argument outlined above. This paradigm or ideology is that of taxonomy: the system of zoological classification which provides us with the concepts of 'species', 'genus', 'family', and so on, and with the system of classification of animals into 'orders' such as 'mammals' and 'reptiles'.

This system of zoological classification is something that we take very much for granted today, to the extent that it has become absorbed in our everyday language. The system was devised some two hundred years ago by the Swedish naturalist Carl Linné, and has subsequently been adopted by the international scientific community. The modern Linnaean system of classification is bound by very strict rules governing the admission of each newly discovered animal and plant to the catalogue of species. The rules covering the naming and description of a new species are every bit as rigorous as those of – say – an International Standards Organisation (ISO) standard for the safety of electrical equipment.

Immense care and attention is given to the smallest detail of nomenclature and its application, so that the system will not fall into disrepute through misuse. Yet despite the enormous utilitarian value of the system in providing a common international language for naturalists, and its great usefulness in cataloguing the plant and animal kingdoms, the system is quite unable to make any contribution whatever to the question of how those kingdoms originated, because the categories with which it deals are the creation of men's minds, not of natural terrestrial processes.

That this is so is far from obvious. Surely the animal called a mouse is so different from the animal called an elephant that only a fool could think the difference exists in men's minds? One is small and white, the other is big and black. There can be no doubt about their difference. The trouble is, however, that it is not in such black and white areas that the Linnaean system is employed by Darwinists, but in the infinite shades of grey in between.

In evolutionist lore the seminal experience for the young Darwin, and the starting point for his theory, was the discovery that the finches and tortoises he found on the Galapagos Islands had evolved from common ancestral stock into different species on different islands in the group because of the varying local conditions on those islands. In the mid-nineteenth century no one objected to Darwin classifying a finch with a long beak on one island as a different species from one with a short beak on another island. Today, many naturalists would reject such differences as amounting to a specific difference and regard it as merely sub-specific variation.

The question of whether types are real or exist only as labels, no more than a by-product of human observation, is an ancient debate stretching back to Plato's time – nominalists versus realists. As far as biology and evolution theory are concerned, the debate remains unsettled because, as Norman Macbeth has pointed out, nature herself capriciously provides evidence for and against both sides. Those who believe species are real can point to examples like the ginko tree which stands in magnificent isolation with no relatives and an unchanging form throughout the geological record. Those who believe species are merely convenient labels point to the willow trees, of which there are countless varieties which blend into each other and present an impossible task to differentiate.[11]

In biology, the debate between realists and nominalists has not been settled but has degenerated into a kind of uneasy truce in which the philosophical issue has been quietly forgotten and replaced by a purely empirical approach.

One of the twentieth century's greatest authorities on taxonomy was Ernst Mayr, Harvard's professor of zoology, whose standard work *Principles of Systematic Zoology* admits that all categories such as 'genus' and 'family' are quite arbitrary in that they seek to describe relationships which cannot be demonstrated experimentally with living populations. Nature is so complex, says Mayr, and so inconsistent, that 'no system of nomenclature and no hierarchy of systematic categories is able to represent adequately the complicated set of interrelationships and divergences found in nature.'[12]

Mayr and his fellow leaders of the synthetic school, Dobzhansky and Simpson, solved the problem by rejecting any attempt to elaborate a theory of biological types, and substituting instead a theory of breeding populations. A population consists of a single species when its members interbreed (producing fertile young) with each other, but not with other such breeding populations.

However, this ingenious solution led to even further complications. The interbreeding criterion means that separate groups may be regarded as different species, even when they appear otherwise identical. This has happened in the case of the fruit fly *Drosophila* where there are a number of interbreeding groups regarded as separate species, even though no one has been able to distinguish them by any other physical characteristic. As well as producing such anomalies, the new definition is inapplicable to large numbers of plants and animals which do not reproduce sexually. And from the evolutionist's viewpoint it is generally of little value because it is usually impossible to determine anything directly about the reproduction of extinct plants and animals simply from their fossil remains.

Whatever the outcome of the debate between nominalists and realists from a philosophical standpoint, biologists and geneticists of all persuasions have rejected taxonomy as anything but a mere convenience in referring to animals and plants. And once the taxonomic categories of 'species', 'genus', 'family' and the like are admitted to be no more than convenient metaphors or contrivances, then we are left simply with a biological realm that consists of individuals – all of which are different.

For example, it might be imagined that all human beings are constituted in exactly the same way, the world over. But surprisingly, this is not the case. Humans vary considerably in matters of detail such as the number of fingers and toes; structure of internal organs like the stomach; the number of bones in the wrist; and other features like amount of body hair and webbed skin between fingers or toes. Usually we marginalise these variations by clinging to a concept of a 'normal' anatomy and dismissing differences as being freaks of nature. In fact, from a genetic standpoint, every organism is unique.

What this means is that the tables drawn up by biologists to classify animals and plants are quite different in character from the tables drawn up by physicists to classify the chemical elements, for example. In the case of the periodic table, each element has uniquely identifying physical properties (the number of particles contained in the atom of that element) and behaves in a predictable way whenever or wherever an experiment is conducted with that element. In the case of the Linnaean system of classification, the majority of living things (and fossils) are distinguished mainly in a statistical rather than absolute way; and sometimes their behaviour is experimentally predictable, sometimes it is not.

This, I believe, is the reason that comparative anatomy must be treated with the greatest care when it is used to infer 'evolutionary' relationships, which actually rely on the taxonomic system rather than blood. The tooth that Gideon Mantell eventually identified by 'coincidence' came not from a true relative of the iguana, but nevertheless, the animal was called 'iguanodon' and the characteristics of the iguana were wished upon the extinct animal. It must have been a reptile; hence it must have been cold-blooded, hence it must have been sluggish; and so on – characteristics which are doubted by some specialists today, such as Robert Bakker of the University of Colorado and Nicholas Hotton of the Smithsonian Institution, both of whom argue that dinosaurs were warm-blooded.[13]

Arguments and theories which rely on drawing attention to evolutionary similarities between 'different species' or even 'different families' thus ultimately depend on maintaining the supposed differences in the first place. To take a rather lighthearted example, one might suppose that the bathroom sponge and the loofah are related, since both are similar aids to scrubbing one's back commonly found in the bathroom. In fact, one is a vegetable which comes from an Indian tree, while the other is an animal which comes from the sea bottom.

The confusion into which some evolutionists have led science by bending the system of zoological classification to fit their theory was pointed out by W. R. Thompson, director of the Commonwealth Institute for Biological Control in Ottawa, who wrote the introduction to a centenary edition of Darwin's *The Origin of Species*:

> The general tendency to eliminate, by means of unverifiable speculations, the limits of the categories nature presents to us, is the inheritance of biology from *The Origin of Species*. To establish the continuity required by theory, historical arguments are invoked, even though historical evidence is lacking. Thus are engendered those fragile towers of hypotheses based on hypotheses, where fact and fiction mingle in an inextricable confusion.[14]

If taxonomy and its handmaiden, comparative anatomy, are misleading in the living world, they are doubly dangerous when applied to fossils. When the zoologist studies the anatomy of living creatures and compares them, he has available evidence not only of the hard parts such as bones and teeth, but also the structure of internal organs, the composition of blood, evidence of skin, hair, coloration and of processes and functions such as body temperature and method of reproduction, none of which survive in fossil form (except in a few freak cases). Often these non-surviving features are crucial in describing an animal, such as whether it is cold-blooded or warm-blooded, whether it bears live young or lays eggs, and the kind of food it lives on. Using taxonomy, evolutionists ascribed reptilian characteristics to dinosaurs. If they were reptiles then they must be cold-blooded. As noted above, this belief is now doubted by some palaeontologists.

Whereas the zoologist can base his description of animals and their living habits on the whole range of their characteristics, the palaeontologist has only the hard parts, bones and teeth, on which to form a judgement. This need not be an insuperable obstacle to accurate diagnosis. Indeed in the field of forensic medicine we are well used to police pathologists performing seeming miracles of detective work in identifying the sex, age and height of dismembered skeletons, and eventually even establishing the identity and cause of death of the victim. In a few cases, such evidence has even led to conviction of the murderer. Professor Keith Simpson solved a case of this kind as a young pathologist during the Second World War. In 1942, workmen demolishing an old Baptist church in south London uncovered a skeleton from which the arms had been cut off below the elbow and the legs removed below the knee. Simpson identified the skeleton as that of a woman by the size of the hip-joints and used 'Pearson's formulae' and 'Rollet's Tables' to estimate the height as 5 feet ½ inch. X-ray photographs of the skull plates showed that the brow plates were completely fused while the top plates were in the process of fusion, which put her age at between 40 and 50. Teeth in the upper jaw had been filled and enabled the dead woman to be identified by dental records. Simpson's detective work placed the dead woman's husband in the dock accused of her murder, and he was convicted.[15]

Similar feats of identification have been performed on human skeletons from remote historical times, such as that of Cleopatra, the wife of Philip II, King of Macedonia, found in 1987. Examples can also be found in palaeontology, where remarkable anatomical descriptions can be given from apparently the most meagre information, especially teeth.

There is, though, a crucial difference between scientific detective work which provides the basis for a case and scientific evidence which is used actually to make a case. Keith Simpson's investigations provided evidence of identification and pointed the finger at murder. From this beginning the police were able to build a case against the husband by uncovering evidence of motive, means and opportunity. Similarly, the Greek archaeologists excavating in the town of Vergina (capital of ancient Macedonia) were able to provide evidence only of the interment of a young woman in her early twenties. It is the circumstances of her burial, such as great treasures, and other circumstantial evidence that points to her being Philip's wife. When a palaeontologist carries out his detective work on the lithified shell of an ammonite, he is compelled to employ intelligent conjecture from the outset. He is using the *circumstances* of the burial as direct evidence of the nature of the extinct creature which used to inhabit the shell. For example, the chalk in which cretaceous ammonites are found is believed to have been deposited in a warm, shallow sea. Thus the creatures found in the chalk will be expected to have been of the type which get a living in such shallow conditions. Obviously, the kind of circumstantial evidence and the kind of conjecture employed by the palaeontologist will depend on the assumptions he has already made about how the rocks were deposited and what pattern of life he expects to see in those rocks.

This kind of detective work is thus almost the opposite of that employed by the forensic scientist. Instead of providing a firm foundation of scientific fact on which the biologist may conjecture the various surrounding circumstances, the palaeontologist who is a convinced Darwinian is providing a basis of conjecture on which the biologist may erect further conjectures – the 'fragile towers of hypothesis on hypothesis' referred to above by Thompson.

If comparative anatomy is unhelpful (or misleading) in seeking to substantiate synthetic evolution, let us turn to one field where Darwinists must feel on absolutely safe ground: the very heart of the synthetic theory, its central mechanism, natural selection.

CHAPTER 11

Survival of the Fittest

When he borrowed the phrase 'survival of the fittest' from
Herbert Spencer, Darwin made it clear that he intended it to
mean precisely the same thing as his own memorable phrase
'natural selection'. 'This preservation of favourable individual
differences and variations, and the destruction of those which
are injurious, I have called Natural Selection, or the Survival of
the Fittest', he wrote in *The Origin of Species*.[1]

The concept of natural selection is fundamental to synthetic
evolution theory. Coupled with random mutation, it is the one
and only mechanism proposed to account for changes in form
fitting a species (sometimes uniquely) to its mode of life – the
streamlining of the dolphin or the giraffe's long neck. According
to Sir Julian Huxley, 'So far as we now know, not only is natural
selection inevitable, not only is it *an* effective agency of evolution, but it is *the* only effective agency of evolution.'[2]

The giraffe has a long neck, according to Darwinists, for three
reasons. First, because an ancestral animal experienced a
mutation which fortuitously gave it a longer neck; second,
because the longer neck gave it some competitive advantage
(such as being able to feed higher up the tree) and thus survive
to produce many offspring; and third, because this natural
advantage also favoured its descendants, a majority of which
would inherit the long neck. The second two stages of this
process are what Darwin meant by his phrase. Darwin also saw
natural selection taking place against the background of a
hostile environment where the majority of offspring die before
reaching maturity or breeding. Darwin's view has thus been
summed up by Simpson in the phrase 'differential mortality' –
the fit leave more offspring than the unfit. The current view is
that a more neutral phrase is preferable and modern synthetic

evolutionists such as Huxley, Mayr and Simpson adopted the phrase 'differential reproduction' as synonymous with natural selection.

As natural selection or differential reproduction is such an important mechanism, one might expect to find a large body of technical literature on the subject, with many detailed studies and observations from the natural world. Regrettably the world's scientific libraries will be searched in vain for such studies because it turns out – for reasons examined in detail a little later – that natural selection cannot be studied in any experimental way. Natural selection means those animals that are best fitted to their environment are the most successful and leave the most offspring. How do we measure the fitness of an animal? By its capacity to survive, say Darwinists. So the fit survive and those who survive are the fittest. Put this way, does natural selection mean anything at all?

Waddington answered this question in 1960 when he wrote:

> Darwin's major contribution was, of course, the suggestion that evolution can be explained by the natural selection of random variations. Natural selection, which was at first considered as though it were a hypothesis that was in need of experimental or observational confirmation, turns out on closer inspection to be a tautology, a statement of an inevitable although previously unrecognised relation. It states that the fittest individuals in a population (defined as those which leave most offspring) will leave most offspring. Once the statement is made, its truth is apparent. This fact in no way reduces the magnitude of Darwin's achievement; only after it was clearly formulated, could biologists realise the enormous power of the principle as a weapon of explanation.[3]

I find it frankly astounding to read a professor of biology describing a tautology as an achievement of any magnitude. However, Waddington's failure to recognise the damaging nature of his admission that 'natural selection' and its synonymous phrase 'survival of the fittest' are nothing more than a harmless tautology is more than made up for by his prescience in accurately foreseeing how Darwinists would use it – as a 'powerful weapon of explanation.'

Darwin conceived his idea of natural selection by analogy with *artificial* selection, something which – as a capable animal breeder himself – he knew a great deal about. He bred pigeons and other animals and travelled extensively in Kent and Sussex discussing animal husbandry with other breeders.

He knew that it was possible for the stockbreeder to change the characteristics of an animal quite substantially in only a few generations – the dairy cow or the sheep, for example. If mankind can change an animal's characteristics by selection, in only a few years, Darwin wondered, what could nature not achieve in many millions of years? Instead of the hand of the stockbreeder, using his experience and his judgement to pick the characteristics he wanted, one had nature herself, acting through the harsh realities of the competitive environment, selecting precisely those characteristics that conferred advantages for continued existence and propagation of offspring, thus ensuring 'the survival of the fittest'.

On first acquaintance, this starkly noble idea appears irreducibly simple. On closer inspection, it is found to be a densely compressed complex of tacit assumptions, few of which correspond with observations of the natural world.

Talk of 'survival' immediately conjures up a lurid vision of competition between the various forms of animal life in a hostile world: competition for the scarce resources of food and living space, of 'nature red in tooth and claw', as Tennyson pictured it for his enthralled Victorian readers. In reality, such competition is very rarely found in nature. There are about 22,000 species of fish, amphibians, reptiles, mammals and birds. In addition there are about 1 million insect species. Some of these thousands of species – notably humans – do compete aggressively, killing competitors for living space and for food. But the species that do are very much in the minority. The overwhelming majority of creatures do not fight, do not kill for food and do not compete aggressively for space in a way that results in the 'loser' dying out.

Earlier in the century it was widely accepted that this kind of behaviour took place with pronounced effects on survival, largely because the evidence was misinterpreted. The male fiddler crab, for instance, has one enormous claw and one normal sized claw which it uses to eat. It was assumed that the enormous claw was to fight its fellow males for the privilege of mating with the most desirable females and/or possessing the most desirable territory. Observation of male fiddler crabs, however, shows that they do not in fact use their large claw to fight. Indeed, they seem to signal the presence of food to their fellow crabs. Far from being a weapon of war, the fearsome claw is an instrument of social co-operation.

There are scores of similar cases where attributes or behaviour were assumed to be aggressive but detailed observation has shown to be nothing of the kind. Fighting between males for 'domination' generally leads to no particular advantage for the 'winner'. His opponent simply goes off elsewhere and mates. Often also, the females will mate as readily with the loser as with the winner. The fighting seems to be much more ritual than actual, rarely resulting in any fatal injury (rather like the fighting of teenage boys in fact). Often when fatal injuries result from intra-species conflict, it is a result of accidents such as when the antlers of male deer become locked, and is fatal to both parties.

The origin of this idea of the struggle for existence resulting in a culling of the less well-adapted was Thomas Malthus's *Essay on Population* published in 1798. Darwin was deeply impressed by Malthus's conclusion that nature regulates the size of human populations automatically through the food supply. He made this Malthusian mechanism the starting point of his theory:

> A struggle for existence inevitably follows from the high rate at which organic beings tend to increase . . . Hence as more individuals are produced than can possibly survive, there must in every case be a struggle for existence, either one individual with another of the same species, or with the individuals of distinct species, or with the physical conditions of life. It is the doctrine of Malthus applied with manifold force to the whole animal and vegetable kingdoms; for in this case there can be no artificial increase of food and no prudential restraint from marriage.[4]

Darwin goes on to give many examples of the nature of 'checks to increase', including the destruction of seedlings by a variety of enemies and the effect of climate on birds' nests (he estimated that the winter of 1854–5 destroyed 80 per cent of the birds in the grounds of his home at Down House in Kent). Although he stressed also that the exact causes of these checks are often obscure.

The key point about this belief from the Darwinist point of view is that it is another example of nature acting blindly. There is nothing, says Darwin, that the animal or plant populations can do to affect the consequences of overpopulation – 'no artificial increase of food and no prudential restraint from marriage'. Death of those not well fitted to exist is the inevitable result.

Today the British Museum of Natural History makes this matter the principal plank in its exhibit on Darwinism. One of the first display cases shows models of rabbits – synonymous with prodigious overbreeding capacity – and explains that the need for rabbits to have a territory, grass to forage, ground for burrows and the need to keep clear of predators limits the available space and hence keeps the rabbit population in check. Similar arguments can be applied to other animals and to plants as well.

These ideas do indeed contain obvious truisms. The question is, do they contain the great principle that Darwinists believe? Consider the example of the fox, more of which now live in towns than in the country. Doubtless it is the instinctive attraction to food behind this change of habitat: but the foxes have not chosen to exercise their ability to change territory before, though presumably they might have done so at any time. This seems to suggest that, unsurprisingly, animals have some element of choice in where they live and are able to move home.

Comparison of animal with human populations is not easy because of the effects of intelligent choice on human behaviour. But nevertheless, it is significant that the average size of the human family in western Europe has fallen dramatically in the last hundred years and has done so as an elective matter, not because of a Malthusian scythe that has blindly cut down excess progeny. It seems to me that at the time Malthus was writing, the effects he observed might have been due to a different cause than a fatal feedback, triggered when population numbers reach a critical limit. When populations expand rapidly, it means that the average size of family increases. At the same time, of course, the number of parents remains the same, so more and more offspring compete for food, care and attention from only a single mother and father, whose resources rapidly become overstretched. What happens as a result is that the infant mortality rate is higher in large families than in small families, where the parents can manage.

It is not the size of population that matters but the size of family in relation to the nurturing capacity of the parents. Malthus was writing in an era of very large families and very high infant mortality. Human population today is declining in western Europe (and infant mortality is low) not as a result of the Malthusian axe but because parents have decided to have small familes: just as foxes have decided to live in towns.

It is perfectly true that many creatures are at risk of losing their lives to hardship and to predators – sometimes even predators of their own species. But again, this form of conflict does not necessarily lead to the kind of competition that would promote a favoured few. The majority of carnivores feed not on prey they have themselves just killed but are scavengers or carrion feeders. This includes legendary hunters such as lions or sharks which frequently eat as a result not of their own direct efforts but those of another lion or shark. (This is also true of mankind.) Thus the 'successful survivor' is not necessarily the most capable hunter-killer and does not necessarily possess the characteristics of such a killer. It follows that if these characteristics are not present, they will not be preserved by breeding.

The Darwinian concept contained another important tacit assumption: that it is within the power of individuals to take action to ensure their survival. That, for instance, the toughest, cleverest, most determined and most enterprising lion will ensure its survival when the prey runs out in its territory by seeking out new territory and new sources of food. But, of course, in many cases it is likely that game has run short through some natural calamity such as drought, fire, or flood. Even if the lion escapes the immediate disaster, there simply may not be alternative sources of food and no action it takes can affect its survival. By the same token, an unenterprising, cowardly, stupid predator in another part of the world may escape the drought or other natural calamity and will survive to breed. Again, it is not the fittest which survives but the luckiest – a quality which is not usually thought of as inheritable.

But it is not merely observation of details that is faulty in the concept of survival of the fittest, it is the concept itself. Why should aggressive competition in which the vanquished fail and die and the victorious survive and prosper be beneficial for the race? As described earlier a whole host of factors, including chance, play a part in the success of an individual of any species. Whether a seed falls on fertile ground or stony ground is a matter of chance and there is no mutation that can assist a sycamore seed to germinate and grow on a wave-washed bare rock.

The concept that the harsh action of the competitive environment is a valuable process that strengthens the breed and weeds out weaklings was a tacit part of the nineteenth-century view of evolution. Nature was a gigantic health club, forcing each species to shape up or ship out. If, in this harsh process, the weak went to the wall, then that was too bad. It is merely nature's way of ensuring that only the fit survive.

The source of this view is nature's indifferent cruelty to the millions of individuals of some species who are born but who perish before attaining maturity and mating. There are only a limited number of winning tickets. Those who fail to grasp such a ticket are doomed to die. Only the toughest and cleverest can wrest a passport to life from nature's cruel grip.

The concept of the struggle for existence, though central to evolution theory in the nineteenth century, has receded in importance; today it is rejected as being either a non-contributory factor in evolution or even actually detrimental to it. Simpson wrote:

> Struggle is sometimes involved, but it usually is not, and when it is, it may even work against rather than toward natural selection. Advantage in differential reproduction is usually a peaceful process in which the concept of struggle is really irrelevant. It more often involves such things as better integration into the ecological situation, maintenance of a balance of nature, more efficient utilization of available food, better care of the young, elimination of intra-group discord (struggles) that might hamper reproduction, exploitation of environmental possibilities that are not the objects of competition or are less effectively exploited by others.'[5]

For Sir Julian Huxley, life's struggle is no more than a banal observation of little significance. 'The struggle for existence,' Huxley wrote in 1963, 'merely signifies that a portion of each generation is bound to die before it can reproduce itself.'[6]

The modern position therefore is that natural selection and the survival of the fittest are no more than harmless tautologies, while the struggle for survival plays no important part in evolution. Where then does this leave synthetic evolutionists from a theoretical standpoint?

According to George Simpson,

> If genetically red-haired parents have, on average, a larger pro-
> portion of children than blondes or brunettes, then evolution will
> be in the direction of red hair. If genetically left-handed parents
> have more children, evolution will be towards left-handedness.
> The characteristics themselves do not directly matter at all. All
> that matters is who leaves more descendants over the genera-
> tions. Natural selection favours fitness only if you define fitness
> as leaving more descendants. In fact geneticists do define it that
> way, which may be confusing to others. To a geneticist, fitness
> has nothing to do with health, strength, good looks or anything
> but effectiveness in breeding.[7]

This sounds solid enough, and it certainly avoids all the old
pitfalls. Natural selection is the process by which the most
successful breeders populate the world, and the less successful
breeders die out – *regardless* of their respective characteristics.
Let us go back to first principles and apply this formula. The
giraffe has a long neck because . . .? Here we get stuck. The only
help we get from synthetic evolution is that the giraffe has
survived because it has survived. Natural selection is unable to
offer any evidence or insight into its evolution because 'the
characteristics themselves do not directly matter at all.'

What this really means is that Darwinists have become reluctant to try and explain any particular characteristic as being
responsible for the giraffe's evolution – even regarding its long
neck – because they would then have to show *how* and *why* that
characteristic has favoured the giraffe over other animals, some
of which are extinct. All that they dare say with impunity is that
the giraffe has survived because it is 'adapted' to its environment – the modern way of expressing the old tautology.

This reluctance on the part of modern Darwinists to comment
on specific characteristics, even very obvious ones, is the result
of two unhappy experiences: first, the battering they received
over the words 'fit' and 'fittest' (which have now been dropped
in favour of 'adapted') and, second, the problems they have had
accounting for the simplest changes. Why for example did the
sabre-tooth tiger become extinct while the common tiger did
not? Why has the modern elephant flourished while many of its
similar relatives became extinct? What was the adaptive
difference?

To summarise, the modern position of the synthetic theory is: the struggle for existence plays no part in evolution. The direction of evolution is determined solely by the characteristics of those animals and plants which are successful breeders. We are unable to say anything of why a particular characteristic might favour – or prejudice – the survival of any particular animal or plant.

Thus 'survival of the fittest', or 'natural selection', or 'differential reproduction', sheds no light on the mechanism of evolution and is only another way of saying some animals live and breed while others die out.

It could reasonably be argued that it is unfair to neo-Darwinists to expect them to have all the answers ready when so many questions in biology remain unanswered. Perhaps instead we should ask them to show us a concrete practical example of natural selection – is there such an example available? Indeed, there is. Every modern textbook with a chapter on selection and evolution, every modern encyclopaedia, contains extended reference to just such an example – the subject of industrial melanism in moths.

The story, as it is usually told, can be summarised as follows. In the first half of the nineteenth century the bark of trees in the Manchester area was progressively darkened by atmospheric pollution from factory chimneys. This gradual darkening affected the peppered moth, *Biston betularia*, which is nocturnal but spends the day resting with wings outspread on the trunks of trees. Before pollution the moth was light grey in colour with dark grey speckles, giving it perfect camouflage against predatory birds, since the tree trunks were also light grey. As the trunks became darker and darker, the moth evolved a darker protective coloration, until by 1898 some 99 per cent of the moths in the Greater Manchester area had the dark coloration.

This phenomenon has been dubbed industrial melanism by Darwinists and is described (for example by the British Museum of Natural History in 1970) as 'the most striking evolutionary change actually witnessed' and as 'demonstrating natural selection'.[8] This story, if true, would be interesting evidence in favour of evolution by natural selection. However, the story turns out to be not quite what it seems.

The change in colour from light grey to dark grey is perfectly real and was recorded over the years by meticulous collecting.

Also, the basic idea of changing colour has been thoroughly tested by H. B. D. Kettlewell at Oxford through experiments under controlled conditions with light and dark moths. The question is, what does the change represent?

Initially, around 1848, only one or a few specimens of the dark variety were collected in the Manchester area. It was assigned the position of a subspecific variety and given the varietal name *carbonaria*. As the tree trunks became darker, the light moths lost their protective camouflage, became conspicuous and fell easy prey to birds. At the same time, the *carbonaria* variety became better and better camouflaged and so began to flourish.

Put more simply: first there were a few dark moths and a lot of light ones. The light ones lost their advantage and the dark ones gained it. All the light moths were eaten, leaving only the dark ones. Far from being an example of evolution or even of natural selection, the peppered moth is an example of a shift in population. The same thing would happen in human terms if some disease were to kill off the white race, but left the black race unharmed. Similar shifts in balance continually occur among animal and plant populations, where one species or variety flourishes at the expense of another. But this process cannot be used to explain the central proposition of neo-Darwinism – how one animal can change into a completely different kind of animal. If 'industrial melanism' is an example of natural selection then Waddington was clearly correct in his assertion that natural selection turns out on close inspection to be merely a tautology.

However, although the story of the peppered moth sheds no light at all on natural selection or evolution, it sheds a great deal of light on how some Darwinists have constructed and promoted their theory, and this is an aspect of the story which I believe deserves closer attention than it has received before. The first part of the story to be examined concerns the original appearance and discovery of the variety *carbonaria*. The claim is sometimes made, or implied, that before the industrial revolution began to blacken trees, the dark grey variety did not exist at all; that it came into being spontaneously when the trees began to darken. *Encyclopaedia Britannica*, for example, says, 'A black mutant of the peppered moth *Biston betularia* was found at Manchester in 1848. Until that time the prevalent form of this species was light grey with dark speckles.'[9]

This statement clearly implies that the dark variety was unknown before 1848. It uses the powerful word 'mutant' to describe the dark form implying both that it was a novel form and was genetically different from the normal type. However, the author (no less an authority than Theodosius Dobzhansky), has left himself a tiny escape hatch in case this claim should be challenged: his use of the word 'prevalent' regarding the light grey variety. 'Prevalent' means having superiority or ascendancy. This can only mean that it has superiority or ascendancy over the black variety – a tacit admission that perhaps the dark variety did already exist after all.

What difference does this make? It makes the difference between being evidence for natural selection and not being evidence for natural selection. If a 'black mutant' suddenly sprang into existence just as the trees darkened and quickly replaced the native population, this would be a powerful example of what every evolutionist since Darwin had been theorising about – a dramatic environmental change being accompanied by an equally dramatic evolutionary change worked through natural selection in combination with mutation. If, however, there have always been some peppered moths with very dark coloration, then a population shift in their favour is what any non-Darwinist would expect.

I said it would be a powerful example if true, but perhaps it would not be so powerful after all. For even if the 'black mutant' did suddenly spring into being, might it not simply be because the dark speckles on its wings were very much bigger than usual? And rather than any kind of novelty, would this not be merely one end of a continuous spectrum from a nearly all light-grey coloration to a nearly all-dark coloration – just another example of sub-specific variation? If so, it could not be used to explain Darwinian evolution, which is trans-specific.

Did the black variety exist before 1848? No one knows. Collectors first noticed the 'melanic' form at around this time. If it did exist before, however, then it would be quite wrong to apply the word mutant to the variety. It could be well within the normal range of variation for that species, just as black men and white men both belong to the species *Homo sapiens* with all varieties of colour in between.

The peppered moth thus brings us to the next most important question to be considered: just how far *can* a species vary?

CHAPTER 12

Green Mice and Blue Genes

Among its many far-reaching effects, France's revolution in 1789 was a revolution in science. It was also one which later would have profound consequences for synthetic evolutionists and their belief in the infinitely plastic and adaptable nature of species. Surprisingly, the originator of the scientific research which illuminates neo-Darwinism from two centuries ago was none other than the Emperor himself, Napoleon Bonaparte.

The story actually begins in Germany in 1747 when Andreas Margraaf showed that various kinds of beet root contain sugar. No one paid any attention to his announcement at the time but in 1786 one of his pupils, Franz Achard, planted beets on his estate at Caulsdorf near Berlin in the hope of finding a commercial source of sugar. Achard's research attracted the attention of King William III of Prussia, who financed the world's first sugar beet factory at Cunern, in Silesia, in 1802. By 1810 the factory was producing sugar successfully on a commercial scale. During this period, of course, France had become embroiled in war with the other European powers. One economically damaging effect of the war was the British naval blockade which prevented imported goods and materials reaching France, and especially the blockade of the West Indies trade routes along which cane sugar was shipped. In 1811, French chemist Benjamin Delessert set up a small factory at Passy and, following the German example, made the first small quantity of crystallised beet sugar. Napoleon was immensely impressed by this scientific achievement and awarded Delessert the prized Légion d'honneur. He also ordered no less than forty factories to be set up in France.

However, now that France had the capability to manufacture beet sugar, it urgently needed to find or breed a type of beet that contained the maximum amount of raw sugar. To achieve this Bonaparte enlisted the greatest botanists in France, through the Académie des Sciences. A programme was begun to breed selectively those sugar beet plants which gave a higher-than-average yield of sugar, a programme which succeeded well. At first the common varieties of sugar beet contained only around 4 per cent sugar on average, but this was rapidly improved – 5 per cent; 10 per cent; 15 per cent. Then things started to go wrong. At 17 per cent average yield, the sugar content of the new plants stuck, and there it has stayed to this day. And, the French discovered, repeated attempts to continue crossing high-yield varieties resulted eventually in the hybrids reverting to the low yields of their ancestral stock. These early geneticists had reached some kind of barrier, but what kind?

Synthetic evolution's most eminent experimental scientist of the twentieth century was Theodosius Dobzhansky, Professor of Zoology at Columbia University from 1940 to 1962 and later at the University of California until the 1970s. Dobzhansky and his co-workers carried out several long and complex series of experiments, making famous the hitherto obscure fruit fly *Drosophila*. This little fly, also called the vinegar fly, is commonly found in many parts of the world and is usually seen buzzing around rotting fruit such as apples. It is of special interest to evolutionists because it is genetically very simple. The genes of every plant and animal species – the blueprint for its offspring coded in DNA molecules – are contained in microscopic bodies called chromosomes which are contained in every living cell. A human being has twenty-three pairs of chromosomes and is genetically complex. *Drosophila* is useful experimentally because it has only four pairs of chromosomes, and can breed a new generation in less than a month.

The types of experiment carried out on *Drosophila* vary in detail but are basically similar. They involve selectively breeding the fly for certain visible characteristics, such as the number of bristles growing on its body or the size of its wings. Harvard's Ernst Mayr has described one such experiment conducted in 1948 which set out to increase the number of bristles in one group, and to decrease the number in a separate group, but starting both groups from the same stock with an average of 36 bristles. By selecting for a lower than normal number of bristles over thirty generations, the experimenters were able to reduce the average carried by the offspring to 25 bristles. After thirty generations, however, the line became sterile and died out. The second group was selected for a higher than average number of bristles and over twenty generations the average rose from 36 to 56. Again, however, sterility became so common that the experiment was wound up.

'Obviously,' says Mayr, 'any drastic improvement under selection must seriously deplete the store of genetic variability.' And, 'The most frequent correlated response of one-sided selection is a drop in general fitness. This plagues virtually every breeding experiment.'[1]

This limit to the amount of genetic variability available in a species Mayr termed 'genetic homeostasis'. It is the natural barrier encountered not only by geneticists attempting to breed fruit flies, and the French botanists attempting to increase the sugar content of the beet root, but by all plant and animal breeders throughout the ages.

Darwin himself, as a breeder of pigeons and other animals, was aware that the amount of variability available was limited. Yet in the first edition of *The Origin of Species* he wrote, 'I can see no difficulty in a race of bears being rendered, by natural selection, more and more aquatic in their habits, with larger and larger mouths, till a creature was produced as monstrous as a whale.'

And although Darwin afterwards thought better of this statement, and removed it from later editions of his book, the substance of its claim nevertheless remains *the* central tenet of synthetic evolution – that bears can become whales, or microbes become elephants by means of random mutation and natural selection. Few Darwinists today could be found to put their names to such a bald manifesto. Yet that is what they believe, or at any rate what they teach in schools and universities.

Darwin's statement is all the more strange because he actually refers to the barrier to variation himself. He quotes Goethe (who had proposed a law of compensation or balance of growth): 'in order to spend on one side, nature is forced to economise on the other side.' And he then adds a few examples of his own: 'It is difficult to get a cow to give much milk and to fatten readily. The same varieties of the cabbage do not yield abundant and nutritious foliage and a copious supply of oil bearing seeds.' Darwin goes on to say that although this law is applicable to animals and plants under domestication, he does not believe it is applicable to species in the wild, although he adds that 'many good observers, more especially botanists, believe in its truth.'

It is very likely that the botanists Darwin had in mind were those (like his close friend Joseph Hooker, Director of the Royal Botanic Gardens at Kew) who would have explained to him that great commercial rewards awaited the first plant breeder to produce a black tulip or a blue rose, but that despite more than two centuries of cross-breeding experiments no one had come close to producing such a variety because they had encountered the same barrier.

The unsuccessful experiments carried out by the French botanists, by the American geneticists and by generations of hopeful Dutch tulip breeders are all concerned with exploiting the natural variability that exists in every animal or plant. All species, according to botany and zoology, exhibit subspecific variation, which is merely the scientific way of expressing the fact that all individuals are different, within certain limits. All the different breeds of dog, for example, from the tiny Chihuahua to the Alsatian, from the Pekinese to the Great Dane are all members of a single species – *Canis familiaris* or the common dog. In exactly the same way, all the races of mankind are members of the same species (indeed, the same sub-species, *Homo sapiens sapiens*), from the pygmies of Borneo to the Zulus of the African plains and from Australian aborigines to professors of genetics.

The amount of natural variation available for cross-breeding is considerable. Dogs have been bred for their speed, for their ability to point at and retrieve game, for guard purposes, as lapdogs and in dozens of directions more or less dictated by their breeders' whims. In cases of extreme selective breeding, such as the Pekinese and even the British bulldog, the variation achieved has been at the expense of other parts of the animals' anatomy. Both these breeds suffer breathing difficulties as a result of facial distortion. Racehorses are also bred selectively to achieve a commercially successful animal and this process too can have an adverse effect on some offspring, often producing animals that are 'temperamental' or too nervous to be ridden.

Stockbreeders and horticulturalists who make their living by breeding desirable strains would have been able to tell the French botanists and the American geneticists the outcome of their breeding trials in advance. Because they know from their own experience that whenever a variation is artificially selected for an animal or plant, to produce some desirable characteristic, the 'improvement' is gained at the expense of some other characteristic of the animal. With domestic animals, the corresponding loss is usually considered unimportant because it affects the ability of the animal to survive if returned to the wild state. The modern dairy cow, for instance, is unable to go even a single day without being milked.

The natural limit on the amount of variation that can be induced in a species is merely the expression of the fact that nowhere in the animal or plant kingdom is there a species that is capable of the infinite biological plasticity demanded by evolution theory, capable of unlimited adaptation to different environments and different modes of life. Living organisms are systems with limited potential for change in which variation of one characteristic reacts on other characteristics, usually with unfavourable results.

This finding is of central importance because it is one that synthetic evolutionists will usually accept, having considered the evidence, but will later on simply forget all about when they are speaking of the Darwinian concept of variation and natural selection. It seems to bring out the Jekyll and Hyde in evolutionists from Darwin down to the present. Darwin withdrew his claim that bears could change into whale-like creatures, yet continued to believe that microbes had evolved into men. Because this is a central issue, it is worth looking in a little more detail at exactly how variation came to occupy the position it does in the synthetic theory.

Darwin began to wonder about speciation among animals when he observed in the Galapagos Islands that the finches and tortoises on each island varied in detail from their counterparts on other islands in the group, while retaining a general similarity. The finches, for example, differed from island to island in terms of the size and shape of their beaks, and the type of food they lived on, but they remained finches. On one island they had strong, thick beaks for cracking nuts and seeds; on another they had smaller beaks and fed on insects; on a third the beak was suited to feeding on fruits and flowers. The acting

Governor of the Islands, an Englishman called Lawson who entertained Darwin to dinner, explained that he could identify the island from which a tortoise came by the shape of its shell: Albermarle Island had a different shell from Chatham, both of which varied from those on James.

Darwin generalised from these and similar observations that animals isolated by geography can change their characteristics over successive generations and adapt to different environments or ecological niches. The mechanism mediating this change he surmised was natural selection – the survival and breeding advantage of those individuals best adapted to their changed environment.

The idea of *natural* selection occurred to Darwin because he had long been interested in how domestic animals and plants are changed by *artificial* selection – by animal husbandry techniques practised by stockbreeders and his fellow pigeon fanciers. The stockbreeder selects for breeding, for example, cattle which produce a high yield of milk and healthy calves, in order to gain those traits which will enable him to build up a dairy herd differing considerably from wild cattle in their milk-producing characteristics. In the same way, Darwin believed, the demands of the environment would act in the stockbreeder's role, favouring the well-adapted and weeding out the badly adapted. At the time, though, Darwin was unable to explain precisely what was the agent of change; for example, what it was that produced a cow yielding higher-than-average milk in the first place. And – more importantly – what specific biological agent caused adaptive changes of character to persist over successive generations. At first he thought the changes might be due to an amplification or magnification of the small natural differences that exist between all individuals of the same species. That there would naturally be a whole spectrum of beak sizes in finches and that on an island where there were only nuts to eat, the large-beaked variety would dominate. However, this explanation was scientifically unacceptable in 1859 because it was wrongly believed that such natural variations were *diluted* by breeding – not amplified.

This was one of the main reasons that Darwinism had almost been consigned to the scientific scrapheap by the beginning of the twentieth century, and equally the reason why the rediscovery of Mendel's experiments in plant breeding rescued Darwinism so comprehensively.

Evolutionists in the early years of the twentieth century realised that with the newly understood laws of genetics, they had a complete explanation for the mechanism of change. To use present-day terminology, the characteristics of each animal and plant are controlled by its genes, the complex chemical structures in its reproductive cells, which are passed on from generation to generation, and which carry the coded instructions which govern the development of the new embryo. Mendel had discovered that some genes dominate, that for example if you cross a short pea plant with a tall pea plant, most of the offspring will be tall. This would explain perfectly how the thick-beaked finch would come to dominate, because his thick beak (if controlled by a dominant gene) would be passed on to a majority of his offspring. Thus the synthetic theory explains Darwin's original observation.

So far, no step in the chain of reasoning has been taken which goes beyond the data. But Darwin's successors felt that the theory as it stood at that point could logically and naturally be extended one further step – and it was a step which appears on the face of it to be not very far beyond the data. If variation and natural selection explained how a finch could change its beak shape to adapt to its island home, and how the giraffe's neck could get longer, then it also explained how one species could turn into a *completely different species.*

After this first intoxicating draught from the tankard of speculation, the newly hatched synthetic evolutionists were brought back to earth with a disillusioning jolt. As we have already seen, ordinary sub-specific variation cannot be pressed into service as the mechanism of evolution for two important reasons: first, because of what Mayr calls 'genetic homeostasis' – the natural barrier beyond which selective breeding will not pass; and, second, because the genetic blueprint for whales is not contained in the *existing* genetic makeup of bears.

Physical characteristics are controlled by genes (or groups of genes acting in concert) and bisexual reproduction ensures that each new individual receives a 'new deal', since the genes of the parents are shuffled together and re-combined like a pack of playing cards. Sometimes, the 'new hand' is very like the old one, as when a human child strongly resembles one parent;

sometimes it is very different. But in every case, the new deal can only be drawn from the existing pack, just as a hand at bridge must contain at least some hearts, or clubs, or diamonds or spades, no matter how much the pack is shuffled.

In terms of physical characteristics, what this means is that genetic re-combination can give rise to variations that are within the range for each species: a finch with a beak a little bigger than before, or a cow that yields more milk than before. What it does not mean is that genetic variation of the ordinary kind is capable of explaining the appearance of entirely novel characteristics. It does not explain the appearance of a wing where before there was only an arm. For the genetic inheritance mechanism is merely one of re-shuffling and re-combination of characteristics already represented in what Dobzhansky called the 'gene-pool' of that species.

This is the specific reason that Dutch tulip growers have never been able to achieve a black tulip or rose breeders a blue rose. There is no gene for black coloration in the gene-pool of the tulip. And, sadly, there are no blue genes for the rose. Ironically neo-Darwinists were rescued from this dilemma by one of the three botanists who had simultaneously re-disco-vered Mendel's pea experiments. Holland's Hugo de Vries had the answer in the phenomenon he believed he had discovered in the evening primrose and which he had dubbed 'mutation'. This was ironic because de Vries believed he was formulating a rival theory to Darwin's. Ultimately, though, de Vries's mutation was absorbed by Darwinian evolution to become the synthetic theory.

This time there seemed no reason to doubt that the theory and the data were in perfect agreement. It was spontaneous, random mutation of the genes that caused novelties to arise; Mendelian genetics that enabled these mutations to be inherited by a majority of offspring; and natural selection that ensured the dominance of the best-adapted species. The fundamental mechanism underlying evolution was therefore chance, acting together with natural selection.

Mutation, as understood by synthetic evolutionists, means the spontaneous change in chemical composition of the genes which occurs quite independently of, and in addition to, the re-shuffling and re-combination that ordinarily occurs in bisexual reproduction. Mutation is today interpreted as being a spontaneous change in the sequence of nucleotides composing the DNA molecules contained in the chromosomes or a spontaneous change in the whole chromosome – a subject looked at in detail in the next chapter. These changes can be brought about by radiation, by chemical agents or simply by copying errors when a cell divides and the DNA double helix separates and replicates itself.

But the most important thing to be understood about mutation is that it is the *only* mechanism proposed by synthetic evolution theory that can account for the appearance of novelty in form. To repeat, ordinary genetic re-combination can account for minor changes in characteristics of an individual – blue eyes rather than brown, tall rather than short – but it cannot ever account for the appearance of any characteristic which is not already contained in the gene-pool of that species.

A breeding pair of white mice might give rise to a brown mouse, providing that the genetic code for brown coloration is already present in the reproductive cells of at least one parent. But they will never, by natural means, give birth to a green mouse, since the genetic code for green coloration is not present in any strain of mouse. The only means by which any entirely novel characteristic can come into being is through the mechanism of genetic mutation – the genetic material in the reproductive cells of one parent must undergo a spontaneous change which must retain the genetic integrity of the original blueprint and hence the viability of the offspring, but which causes some change in form.

The only way a bear can become a whale is through mutation. No amount of natural selection alone will do it, as Darwin was at first inclined to think.

'It must not be forgotten,' says Ernst Mayr, 'that mutation is the ultimate source of all genetic variation found in natural populations and the only new material available for natural selection to work on.'[2]

Neo-Darwinists thus have two completely separate mechanisms at their disposal to account for changes in the plant and animal worlds. To account for sub-specific change – such as the Galapagos finches – they have ordinary variation coupled with natural selection. Above the species level (and this is the only kind of evolution that really matters) they have spontaneous genetic mutation coupled with natural selection.

Since sub-specific change is trivial, and since 'natural selection' is a harmless tautology which cannot alone explain evolution, it is pretty clear that the whole theory rests finally upon the phenomenon of 'spontaneous genetic mutation'. It is to this mysterious matter that we now turn.

CHAPTER 13

Of Cabbages and Kings

The instructions for reproduction of a cabbage or a king are contained in dark thread-like strands in the nucleus of its cells, called chromosomes, which are long molecules of DNA and protein. The instructions themselves consist of sequences of four chemical groups, called nucleotides, strung like clothes on a washing line and identified by their initial letters – C, T, A, and G (cytosine, thymine, adenine and guanine).

The genetic meaning of each sequence – the kind of physical characteristic it gives its owner – depends on the sequence of nucleotides (C, T, A and G) and the position of that sequence on the 'washing line'. The sequence TCA for example is the genetic code for an amino acid called serine which is important in building membranes. The sequence CCA is the code which causes the synthesis of an amino acid called proline which is widely used in building connective tissues.

The total number of instructions, or sequences, in an organism as complex as mankind not surprisingly runs into millions – too many in fact for one chromosome. Each chromosome is like a magnetic tape; but there are so many instructions they run over onto another tape, and another. In man there are 23 chromosomes in the 'tape library'. In the fruit fly there are 4, while the humble land snail has 27 and, curiously, the simple goldfish has 47. When she wrote that a rose is a rose is a rose, Gertrude Stein must have been unaware that roses exist with 14, 21, 28, 35 and 56 chromosomes.

However, despite the genetic code taking up so 'much' room (actually a chromosome is only one-hundredth of a millimetre long) some 90 per cent of the recording space in the tape library is empty. The genetic instructions which actually cause the manufacture of proteins occupy only 10 per cent of the available coding space. Or to be more exact, only 10 per cent of the sequences in the chromosomes cause anything to be replicated. The function of the other 90 per cent of sequences is unknown at present. Although it does not replicate, it may affect the positioning (and hence the genetic meaning) of the sections that do replicate. This mysterious 'unused' 90 per cent of genetic coding space is very important to neo-Darwinists for reasons explained later on.

While enough is known of the genetic code to justify saying that geneticists understand it in principle, much remains mysterious. Unfortunately, it is not a simple case of a single gene, at a single location, controlling a single characteristic. Various locations are linked together to control groups of characteristics in a non-obvious way.

When a male sperm unites with a female egg in bisexual reproduction, the DNA molecules of each split apart, like a zip unzipping, and cross-join to form the set of chromosomes of the first cell of the new individual. Also, when the first cell divides and re-divides to form the new embryo, the DNA molecules unzip and replicate themselves exactly, thus perpetuating the blueprint or building instructions in every cell of the new individual both for its embryonic form and for life.

A number of things can go wrong with genetic reproduction. Whole chromosomes can sometimes be replicated wrongly and individual genes (nucleotide sequences) can be replicated incorrectly, either because of faulty selection of nucleotides or because the right sequence is put in the wrong location, perhaps shifted one position to the left or right. Faulty replication of a sequence will result in the issuing of instructions to the factory to make the wrong product. If the sequence TCA is wrongly replicated as CCA (just like a typing error in a document) then the cell will manufacture proline instead of serine, with unpredictable consequences.

The Darwinist interpretation of these discoveries is that the genetic mutation which leads to novelties in form is caused by spontaneous alterations in the DNA molecule and hence of the genetic code. A good many of these genetic changes happen without actually altering the physical characteristics of the individual who carries them. These genes are called recessive, latent or inhibited and it is believed by Darwinists that they are 'stored' in the 90 per cent of unused genetic material. Under certain circumstances these latent genes can replace the usual dominant genes expressed in the physical characteristics of the offspring.

This is a very important mechanism for neo-Darwinism, for without some kind of storage of novelties 'waiting to happen' as it were, it would be expecting too much of evolution to demand that it not only produce novelties spontaneously but also produce them exactly *when* needed because of some ecological crisis.

Evolution, according to neo-Darwinists, is due basically to copying errors. Although the DNA is astonishingly stable from generation to generation, and although reproduction is error-free to a far higher degree than the most efficient man-made copying systems, there are occasional mistakes: an A instead of a T or a G instead of C. Or position 51 or 53 instead of 52.

These copying errors can happen spontaneously, or they can be caused by some outside mutagenic agency such as radiation or highly toxic chemicals, like mustard gas. Ultra-violet light from the sun is mutagenic, but has very little penetrating power and hardly gets beyond the skin. X-rays on the other hand penetrate deep into the human body causing much direct cell damage and damaging DNA which will continue to replicate in a faulty way.

The results of such copying errors are tragically familiar. In body cells, faulty replication shows itself as cancer. Sunlight's mutagenic power causes skin cancer; the cigarette's mutagenic power causes lung cancer. In sexual cells, faulty reproduction of whole chromosome number 21 results in a child with Down's syndrome. Only 'germinal' mutation – that is, the mutation of sexual cells in the male sperm or female egg – can result in an inheritable variation, believe Darwinists. According to the same theory, 'somatic' or body-cell mutation cannot be inherited, and this is the specific reason that Darwinists are also anti-Lamarck-ian: they say that even if an animal's mode of life should result in somehow bringing about mutation in the creature's body cells, there is no mechanism for these changes to be passed on to the next generation; only sexual cells do that job, not body cells.

From the point of view of assessing whether neo-Darwinists have made their case, the key issue in microbiology is the rate of mutation. This has to be frequent enough to provide a realisti-cally probable occurrence of novelties, but not so frequent that no two generations are ever the same and evolution runs haywire. Too little and we are stuck in the primeval ocean unable to set the evolutionary ball rolling; too much and we are living in an unstable nightmare world of monsters.

The question of just how much mutation takes place is thus of considerable importance, and is a much studied subject. What conclusions have Darwinists come to?

Sir Julian Huxley estimated that the rate of inheritable mutation was around one in every million births.[1] French micro-biologist Jacques Monod has estimated the rate at one in ten thousand births.[2] The reason for this diversity of opinion between the Professor of Zoology at King's College, London, and the Director of Paris's Pasteur Institute is simple. It is because the beneficial spontaneous genetic mutation remains no more than a hypothetical necessity to the neo-Darwinist theory. No one has ever observed a spontaneous inheritable genetic mutation that resulted in a changed physical character-istic, aside, that is, from a small group of well-known and usually fatal genetic defects. Because no one has ever observed such an event, no one really knows whether they occur at all and if so, how often. Because deleterious mutations are known to occur, neo-Darwinists appeal to the statistics of large numbers. If deleterious mutations can occur, then given enough time beneficial mutations can occur. There is no evidence for this claim. But it is irrefutable.

'Detectable results of germinal mutation among people are only very rarely encountered,' says the author of the 1984 *Encyclopaedia Britannica* entry on Human Genetics. 'Thus the actual rate of mutation in human chromosomes defies full measurement. Efforts to measure mutation rate therefore are most conveniently directed towards selected dominant . . . mutations for which [physical] recognition is easier; indirect (inferential) methods of measurement are still required.'[3]

The dominant mutations where physical recognition is easier include achondroplasia (dwarfism), Huntington's Chorea and Down's syndrome. And it is on deleterious mutations of this kind that the human mutation rate is estimated. *Encyclopaedia Britannica* cites a general mutation rate for human genes of 4 mutations per 100,000 gametes (that is, male sperm or female egg).

This rate of mutation sounds impressively high. But the

reality is very different. What has happened is that the rate of mutation has been inflated by the simple device of making the definition of the term 'mutation' so elastic that it can include any and every inheritable change – including those that invariably lead to fatal diseases. Even though Darwinists are perfectly aware that no individual can ever benefit from dwarfism, they include achondroplasia as a genetic mutation, and Huntington's Chorea, and neurofibromatosis, and put them all towards measuring the rate of mutation.

They also lump in other types of mutation, many of which are caused by natural radiation and other mutagens, and which again can only result in exclusively harmful results for the offspring. It is obvious that, in reality, a major proportion (perhaps all) of these so called mutations cannot possibly lead to any benefit and are bound to fail. Only a tiny fraction have any prospect of turning into a novelty of form that might conceivably be helpful, and even the existence of this tiny minority is granted only because the claim that they might exist cannot be refuted.

The fact is that more than 99 per cent of so called mutations should not be included in the measured rate but these mutations cannot be positively excluded because no one can predict which mutations will be useful and which will not. It is conceivable, after all, that chemical pollution might produce a useful variation, or that dwarfism might be adaptive if the circumstances of the environment changed.

I believe that Darwinists should screw up their courage and come clean by separating the two cases, reserving the term mutation for any change in genetic coding, whatever the cause and whatever the effect, and use some other term – perhaps 'novation' (novelty-producing mutation) – to describe the kind of mutation they say is potentially useful. Note that 'novations' do not have to exclude all copying errors or blunders – only those whose genetic consequences are already known not to lead to evolutionary novelty, such as Huntington's Chorea.

Can we estimate the rate of 'novation' as opposed to the rate of mutation? Yes we can. The rate of novation is a number that is vanishingly small (if not actually zero). It is a number so small that in order to account for synthetic evolution by random mutation one has to have an almost religious faith in the power of extremely unlikely events and very long time scales.

Thus at the very heart of the synthetic theory of evolution is a single, central matter: improbability. How we deal with this question alone either convinces us of the validity of neo-Darwinism or convinces us of its impossibility. Regardless of evidence from all other sources – geology, stratigraphy, palaeontology, comparative anatomy, zoology, botany and genetic studies – it is the question of the probability of life coming into being spontaneously and evolving spontaneously – without outside assistance – that separates the sheep from the goats.

The two camps might justly be represented by respective champions who are almost always quoted when the issue of mutation and natural selection is debated: William Paley, the eighteenth-century Archdeacon of Carlisle, and Sir Ronald Fisher, founder of the modern mathematical school of genetic studies.

Paley, in his influential book *Natural Theology* published in 1828, observes that if, while out walking, you were to find lying on the ground a watch full of intricate mechanisms you would have to conclude that it had been wrought by a creator; it would be impossible to believe such a machine had come into being accidentally. The human body is infinitely more complex and intricate than any watch mechanism, so we must conclude that it too has a purposeful creator.[4]

Fisher, whose equally influential book *The Genetical Theory of Natural Selection* appeared in 1930, observed quite simply that natural selection is a mechanism for generating improbability.[5]

Paley's watch – the argument from design – is not really a serious scientific argument and can be easily refuted. Most recently it has been very effectively dismissed by zoologist Richard Dawkins in his book *The Blind Watchmaker* where he points out simply that we are not obliged to see the hand of God in such seeming miracles and that one individual's inability to conceive of highly improbable events does not make those events impossible. Those who employ Paley's argument, says Dawkins, should speak only for themselves.[6]

Dawkins' own way of dealing with the improbability of evolution by mutation, however, a way that is representative of the modern neo-Darwinist view, makes use of a fallacy many times more subtle than anything Paley dreamed of on his horological rambles. One has to observe each step of Dawkins' argument very carefully to spot exactly where the fallacy comes in.

To get from a barren primeval Earth to a complex organ like the human eye in a single step, says Dawkins, would require random spontaneous events that are so improbable as to be practically impossible. However, he says, it is not so wildly improbable to get there in a series of small steps, each step requiring admittedly improbable events, but not so improbable as to be practically impossible. And, of course, vast ages are available in the geological past for these smaller steps to be accumulated. Adding up a long series of small, improbable but not impossible steps can cumulatively give rise to a complex mechanism such as the eye which *overall* is of incredible improbability.

More simply, you can get a result whose improbability is so great as to be practically impossible by adding together a lot of little steps whose improbability is high, but nevertheless practically possible. Moreover, says Dawkins, if you break up the process into steps that are cumulative, it is quite likely that you only have to contend with one or a few steps of extremely low probability – those at the beginning – and once the evolutionary ball is rolling, the events required become less and less improbable.

'My personal feeling,' he says,

> is that, once cumulative selection has got itself properly started, we need to postulate only a relatively small amount of luck in the subsequent evolution of life and intelligence. Cumulative selection, once it has begun, seems to me powerful enough to make the evolution of intelligence probable if not inevitable. This means that we can, if we want to, spend virtually our entire ration of postulatable luck in one big throw, in our theory of the origin of life on a planet.[7]

Dawkins' argument is a modern rendition of the traditional Darwinist approach and the error it falls into is that dubbed the 'Statistical Fallacy' by Francis Crick.[8] Although employing modern language it is really in principle the same as Darwin's own claim that:

> Slow though the process of selection may be, if feeble man can do so much by his powers of artificial selection, I can see no limit to

the amount of change, to the beauty and infinite complexity of the co-adaptations between all organic beings, one with another and with their physical conditions of life, which may be effected in the long course of time by nature's power of selection.[9]

If Paley's watch is the argument from design, then the Darwinian case might be called the argument from probability. What does it really amount to?

Suppose we have a highly improbable event such as a perfect deal in bridge, where each of the four players receives a complete suit of cards. The odds against this happening are billions of billions of billions to one. Let us assume that since being manufactured the cards have been used for 99 deals and on the 100th time the pack was shuffled, the perfect deal arose. Can we say that each of these previous shuffles, deals and plays of hands (number 1 for instance) was a cumulative event that ultimately contributed to the perfect deal? Can we reduce the ultimate odds against the perfect deal by attempting to spread them around more thinly between the intermediate steps? Not *afterwards*, note, when we know the result, but at the time each step is occurring?

The answer is no, we cannot. Like the supposedly evolving DNA, the cards have a memory in that the previous deals have contributed to their current order and the ultimate perfect deal. But being part way towards a perfect deal does not alter the odds on the ultimate deal, because some of the key random events determining the ultimate outcome have not yet taken place.

The same is true of Dawkins' hypothetical evolutionary model. Although the earlier steps in his evolution process are seen retrospectively to contribute to the end result, that does not affect the probability of each intermediate step coming about *at the time*. It is perfectly true that the minimum overall probability we have to deal with in considering the evolution of a human eye is a product of all the probabilities of the individual steps necessary to attaining that end – but paradoxically, this does not diminish the probability of each individual step when the need for the correct sequence is also taken into account.

What Dawkins is saying with his cumulative evolution argument is that the probability of each single step in a cumulative process must be *less* than the whole probability of leaping

GEOLOGICAL TIME-SCALE

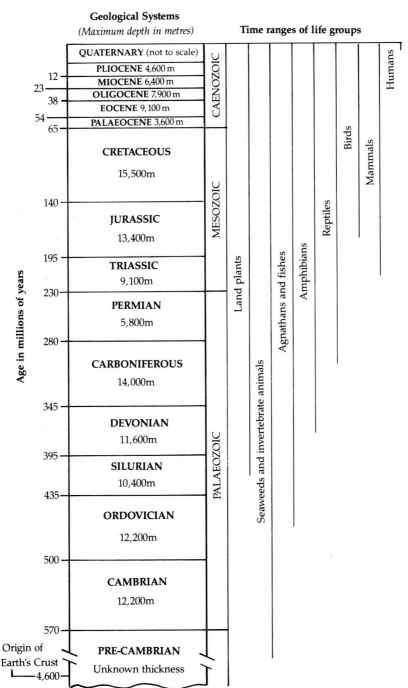

Geological Systems
(Maximum depth in metres)

Time ranges of life groups

Age in millions of years	Geological System (Maximum depth in metres)	Era
	QUATERNARY (not to scale)	CAENOZOIC
12	PLIOCENE 4,600 m	
23	MIOCENE 6,400 m	
38	OLIGOCENE 7,900 m	
54	EOCENE 9,100 m	
65	PALAEOCENE 3,600 m	
	CRETACEOUS 15,500m	MESOZOIC
140	JURASSIC 13,400m	
195	TRIASSIC 9,100m	
230	PERMIAN 5,800m	PALAEOZOIC
280	CARBONIFEROUS 14,000m	
345	DEVONIAN 11,600m	
395	SILURIAN 10,400m	
435	ORDOVICIAN 12,200m	
500	CAMBRIAN 12,200m	
570		
Origin of Earth's Crust —4,600—	PRE-CAMBRIAN Unknown thickness	

Life groups: Land plants, Seaweeds and invertebrate animals, Agnathans and fishes, Amphibians, Reptiles, Birds, Mammals, Humans

The geological column of accepted dates for sedimentary rocks forming most of the Earth's crust. Notice the slow rate of sedimentation (averaging 0.2 millimetres per year). The column is claimed to be dated by radioactive methods but most sedimentary rocks do not contain radioactive materials. (*Natural History Museum*)

Sedimentary rocks are said by uniformitarians to take millions of years to form at slow rates of deposition. But full-size trees found in position of growth point to rapid burial. This fossil tree excavated from Carboniferous rocks near Edinburgh stands in the grounds of the Natural History Museum. (*Photo*: *author*)

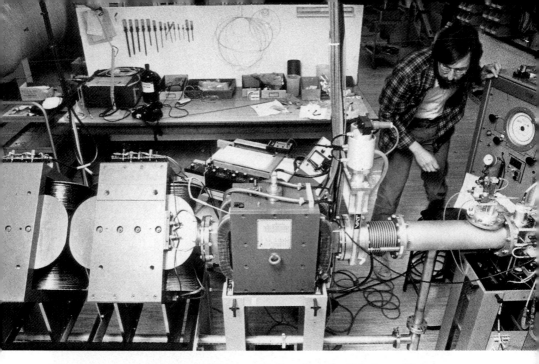

Oxford University's carbon-dating laboratory. Radiocarbon should be in equilibrium in the atmosphere but measurements show production exceeds decay by up to 38 per cent. This could mean the atmosphere is less than 30,000 years old.
(*Photo: Science Photo Library*)

The Earth's radioactive clocks are running fast. Neutrino particles penetrating the Earth's crust in historical times have speeded up the rate of decay of all radioactive isotopes used for dating. Supernovae, such as the Crab nebula shown here, shower the Earth with neutrinos. Hundreds of such stellar explosions have been mapped by astronomers.
(*Photo: Science Photo Library*)

Mount Kilauea in Hawaii is an active volcano. Geologists from the Hawaiian Institute of Geophysics used the Potassium-Argon method to date volcanic lavas and found ages ranging from 160 million years to 3 billion years. In fact the rocks are known to be modern – only 190 years old.
(*Photo: Science Photo Library*)

Fourteen million tons of meteoric dust enters the Earth's atmosphere each year. If the Earth were 4,600 million years old, there would be a layer of dust 180 feet thick over the surface, and a comparable amount on the Moon. In fact there is merely an inch or two of dust on the Moon as this astronaut's footprint shows.
(*Photo: Science Photo Library*)

Archaeopteryx is said by Darwinists to have evolved from dinosaurs called 'coelosaurs' and to be ancestral to birds. But coelosaurs did not have collar bones while *Archaeopteryx* does. And while birds' wings are composed of the 2nd, 3rd and 4th fingers of the hand, *Archaeopteryx*'s wing is composed of the 1st, 2nd and 3rd fingers.
(*Photo: Science Photo Library*)

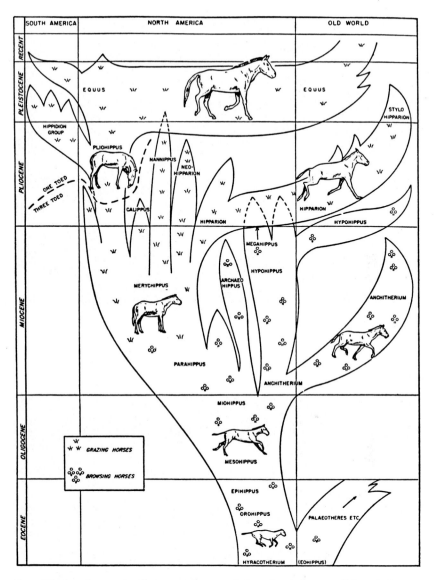

One of the best known 'evolutionary sequences', that of early horses. Although the diagram shows an unbroken line, there are major gaps in the sequence, for example between *Eohippus* and its supposed ancestor, and between *Eohippus* and its supposed descendant, *Miohippus*. (*Drawing from* Horses *by George G. Simpson*)

Fossil ammonites from the lias of Blockley. *Liparoceras* (left) and *Androgynoceras* (right). Darwinists have variously claimed that the species on the left is the ancestor of the one on the right; that the species on the right is the ancestor of the one on the left; and that the two forms are male and female of the same species. Darwinist theory can accommodate all three conclusions. (*Photo: author's collection*)

Fossil ammonites, from the gault clay of Kent. (Top) *Douvilleiceras, Beudanticeras, Hoplites*. (Bottom) *Euhoplites, Anahoplites, Dimorphoplites* and *Mortoniceras*. Darwinists believe they are an evolutionary sequence but there are no intermediate species in the beds between. (*Photo: author's collection*)

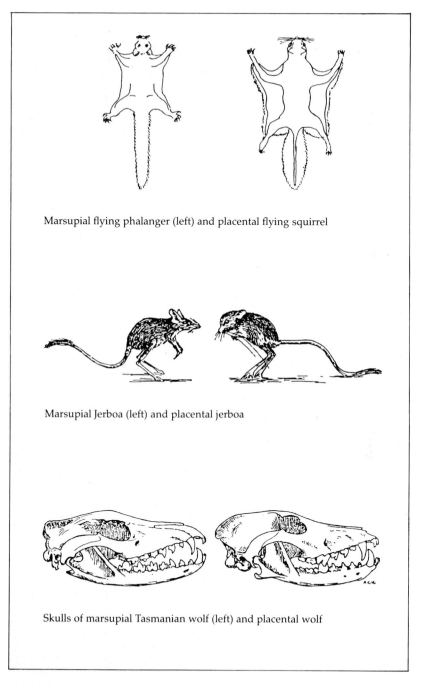

Marsupial flying phalanger (left) and placental flying squirrel

Marsupial Jerboa (left) and placental jerboa

Skulls of marsupial Tasmanian wolf (left) and placental wolf

The Australasian marsupials on the left are among many species identical with their placental counterparts. Darwinists believe these species have evolved in parallel but quite separately over the last 65 million years from a common shrew-like ancestor. By random mutation? (*From* The Living Stream *by Sir Alister Hardy*)

Restoration of 'Piltdown man'.

Restoration of *Pithecanthropus erectus*.

Darwinist restorations based on fragmentary finds of bones and teeth always manage to convey a distinctly 'missing link' quality to their former owners. The convincing Piltdown man is wrongly based on a simple forgery associating a normal human skull with the jaw of an ape. But the same 'artistic skill' and imagination have been applied to the genuine fossils. (*Restorations by J. H. McGregor. From* Men of the Old Stone Age *by Henry Fairfield Osborn*)

Restoration of Neanderthal man.

straight to the finished result, simply because each step itself is less than the whole. But this is simply wrong. The improbability of step number 2 correctly following step number 1, correctly followed by step number 3 and so on for 100 mutations, is as great as leaping to the 100th step in one go.

What is more, the greater the number of steps into which we break up the overall leap, the more improbable it becomes that they will all take place in the right order. Mutation number one might be the first step in evolving an eye (or magnetic detector or infra-red detector or X-ray detector). But the probability of the next mutation step affecting that organ being the second step needed for an eye is not increased thereby. It does not become any easier for an eye to come into being just because the first of the 100 or 1,000 accidents needed has taken place, *even if that first step is a very important general innovation such as light-sensitive tissue.*

Modern Darwinists seem to have a profoundly optimistic belief that the occurrence at an early stage in evolution of such a fundamental innovation – cells which are sensitive to light – makes cumulative selection of vision somehow less improbable. But the existence of light-sensitive tissue has no effect whatever on the probability of the mutation of a lens, or an iris mechanism or an eyelid or anything else.

Of the vast range of characteristics spelled out by DNA, the next copying error is more likely to be about something else entirely – the beginnings of a wing or lung perhaps – or it may be the wrong step, such as providing eyelids before providing the muscles to move them, thus blinding their possessor.

Darwinists say that this case cannot be made against them because purpose has no place in their argument. The Darwinist mechanism of evolution is blind; and its outcome is arbitrary. Neither nature nor Darwinists care what the end result of the selection process is: the species that inherit the earth will simply be those that nature has blindly selected and are best adapted to their habitats and way of life. There is no 'perfect deal' in evolution, say Darwinists; no final result to be anticipated. There is only an infinity of uncertainty leading on always to novelty dictated only by changed environmental circumstances.

This apparent rejection of purpose is deceptive. Darwinists are very firmly convinced that there is a predictable and desirable end result for any given habitat or way of life. Indeed, that is the very origin of the Darwinian concept. Any individual less than perfectly adapted will ultimately be replaced by another individual that is better adapted, unless it has no competitors, in which case it will ultimately acquire such competitors and be displaced. This process will inevitably continue until eventually it will reach a conclusion where the process cannot take place any further.

Although the mutation part of this process is random it is clear that the selection part very definitely is not random: in fact it is keyed perfectly to the template of ecology and habitat. Otherwise why would camels be able to survive for days without water? Or sea otters hold their breath for long periods? Darwinists use this very argument to account for the parallel evolution in isolated environments of identical animals, such as Tasmanian marsupial wolves and European placental wolves.

Given any specific habitat, and any existing set of animal characteristics, it is possible in principle to set down precisely what characters that animal will have to acquire in order to be perfectly adapted – and thus irreplaceable. In theory it ought to be possible to use one of the Darwinists' favourite computer-based genetic software systems to show in live animation exactly how that creature should mutate to get most efficiently from its current position to the theoretically perfect position.

This idea is not merely conjecture. Darwinists have already done this very trick in the case of the extinct dinosaur. In 1982, Dale Russell and R. Seguin of Ottawa University published a paper describing the partial skeleton of the dinosaur *Stenonycho-saurus* which had been found in Alberta in 1967. Their paper covered the work involved in building a flesh and bone reconstruction of this species. However, Russell and Seguin decided to take their reconstruction one step further. Because *Stenony-chosaurus* was small (about 6.5 feet tall) and a biped with flexible fingers and a relatively large brain, the authors asked what would have happened if the creature had continued evolving in a Darwinian fashion to the present day, instead of becoming extinct. Their resulting reconstruction looks astoundingly human-like apart from a certain unfortunate reptilian glare.[10]

Russell and Seguin's reconstruction is, of course, merely an imaginative thought experiment, carried through to be entertaining and thought provoking. But I believe it represents quite fairly the belief shared by most people – Darwinists and non-Darwinists alike – that there is an inevitability about the design of man and of all other species. There is a beauty and grace in the flight of the bird which less developed flying designs do not possess and which enables birds not only to conquer the air but to dominate it.

This perfect fitness of form is also evident in the improving design of human artefacts such as the car and the jet airliner: decades of experience with a plurality of trial designs passing through the filter of experience into a single optimum design – a process frequently referred to as design evolution (though not, of course, happening by chance).

Darwinists should steel themselves to recognise that the flight of the eagle and the sprint of the cheetah represent a 'perfect deal' for evolution, an end-result that cannot be improved on. These animals have not arrived at an arbitrary point; they have arrived at the point that uniquely positions them to exploit their habitats.

Throughout the lifetime of the Earth, there always was a definable probability of them getting to that end point by chance alone – and that probability always was vanishingly small.

Simpson's claim, quoted in Chapter 11 ('The characteristics themselves do not directly matter at all. All that matters is who leaves more descendants over the generations. Natural selection favours fitness only if you define fitness as leaving more descendants'), simply will not do. It is part of the alibi Darwinists have prepared in case they are asked to comment analytically about the inherited characteristics of any given animal or plant.

To illustrate his theory of cumulative steps, Dawkins wrote a computer program which randomly generates symmetrical figures from dots and lines. These figures, to a human eye, have a resemblance to a variety of objects. Dawkins names some of them as insects and animals such as bat, spider, fox, caddis fly. Others he gives names like lunar lander, precision balance, spitfire, lamp and crossed sabres.

Dawkins calls these creations 'biomorphs', meaning life shapes or living shapes, a term he borrows from fellow zoologist Desmond Morris. He also feels very strongly that in using a computer program to create them he is in some way simulating evolution itself. His approach can be understood from this extract:

> Nothing in my biologist's intuition, nothing in my 20 years experience of programming computers, and nothing in my wildest dreams, prepared me for what actually emerged on the screen. I can't remember exactly when in the sequence it first began to dawn on me that an evolved resemblance to something like an insect was possible. With a wild surmise, I began to breed generation after generation, from whichever child looked most like an insect. My incredulity grew in parallel with the evolving resemblance . . . Admittedly they have eight legs like a spider, instead of six like an insect, but even so! I still cannot conceal from you my feeling of exultation as I first watched these exquisite creatures emerging before my eyes.

Dawkins' book represents the latest generation of neo-Darwinist belief quite fairly, and this experiment to my mind exemplifies perfectly the nature of that belief. The fact that his experiment was carried out by computer also gives it a certain authority in the minds of some people who are not themselves highly computer literate.

Dawkins not only calls his computer drawings 'biomorphs', he gives some of them the names of living creatures. He also refers to them as 'quasi-biological' forms and in a moment of excitement calls them exquisite creatures. He plainly believes that in some way they correspond to the real world of living animals and insects. But they do not correspond *in any way at all* with living things, except the purely trivial way that he sees some resemblance in their shapes. The only thing about the 'biomorphs' that is biological is Richard Dawkins, their creator. As far as the 'spitfire' and the 'lunar lander' are concerned there is not even a fancied biological resemblance.

The program he wrote and the computer he used have no analogue at all in the real biological world. Indeed if he set out to create an experiment that simulates evolution, he has only succeeded in making one that simulates special creation, with himself in the omnipotent role.

His program is not a true representation of random mutation coupled with natural selection. On the contrary it is dependent on artificial selection in which he controls the rate of occurrence of mutations. His biomorphs show no real novelty arising even within the context of Dawkins' own imaginative interpretations. There are no cases of bears turning into whales.

There is also no failure in his program: his biomorphs are not subject to fatal consequences of mutations like real living things. And, most important of all, he chooses which are the lucky individuals to receive the next mutation – it is not decided by fate – and of course he chooses the most promising ones ('I began to breed . . . from whichever child looked most like an insect'). That is why they have ended up looking like recognisable images from his memory. If it were really decided by chance, Dawkins would still be sitting in front of his screen watching a small dot and waiting for it do something.

In fact, his experiment shows very much the same sort of results that field work in biology and zoology has shown for the past hundred years: there is no evidence for beneficial spontaneous genetic mutation; there is no evidence for natural selection (except with a trivial meaning); there is no conclusive evidence for evolution. There is only evidence of an unquenchable optimism among Darwinists that given enough time, anything can happen – the argument from probability.

It is interesting, though, that Dawkins should choose a computer program to attempt to simulate evolution, because there is at least one very important way in which computer programs resemble genetic processes. Each instruction in a program must be carefully considered by the programmer as to both its immediate effect on the computer hardware and its effects on other parts of the program. The letters and numbers which the programmer uses to write the instructions have to be written down with absolute precision with regard to the vocabulary and syntax of the programming language he uses in order for the computer system to function at all, and even the most trivial error can lead to a complete malfunction. In 1977, for example, an attempt by NASA to launch a weather satellite from Cape Canaveral ended in disaster when the launch vehicle went off course shortly after take-off and had to be destroyed. Subsequent investigation by NASA engineers found that the accident was caused by failure of the on-board computer guidance system – because a single comma had been misplaced in the guidance program.

Anyone who has programmed a computer to perform the simplest task in the simplest language – BASIC, for instance –

will understand the problem. If you make the simplest error in syntax, misplacing a letter, a punctuation mark or even a space, the program will not run at all.

In just the same way, each nucleotide has to be 'written' in precisely the correct order and in precisely the correct location in the DNA molecule for the offspring to remain viable, and, as described earlier, major functional disorders in humans animals and plants are caused by the loss or displacement of a single DNA molecule, or even a single nucleotide within that molecule.

In order to simulate neo-Darwinist evolution on his computer, it is not necessary for Dr Dawkins to devise complex programs that seek to simulate life. All he has to do is to write a program (containing a largish number of instructions) that continually regenerates its own program code, but randomly interferes with the code in trivial ways, such as transposing, shifting or missing characters. (The system must be set to restart itself after each fatal 'birth'.) The result of this experiment would be positive if the system ever turns out to have developed a novel function that was not present in the original programming. One way of defining 'novelty' would be to design the program so that its sole function initially was to replicate itself. A novel function would then be anything other than mere reproduction. In practice, however, I do not expect the difficulty of defining what constitutes a novelty to pose any problem. For it is extremely improbable that Dr Dawkins' program will ever work again after the first generation, just as in real life mutations cause fatal genetic defects, not improvements.

Of all the difficulties facing neo-Darwinism, the improbability of spontaneous genetic mutation leading to beneficial novelties in form ought to be the major source of concern. This is so because, as explained in the previous chapter, it is the one and only source of inheritable variation above the species level available – the ordinary variation caused by genetic recombination not being capable of achieving novelties above the species level.

In practice, some synthetic evolutionists get round the problem in a number of ways, two of which we have already seen: they pretend many more mutations occur than actually take place, by including fatal genetic defects, and they pretend that splitting up the overall evolution process of a complex organ like the eye somehow reduces the improbability of those separate steps coming about by accident in the correct sequence.

In addition there are two further important devices sometimes used by Darwinists that can be found in a variety of guises. The first is to ignore the difficulties inherent in genetic mutation and 'fudge' it together with ordinary genetic variation – ignoring also the fact that genetic re-combination alone cannot give rise to novelties in form above the species level. This is the 'industrial melanism in moths' fudge, for instance. A second device is to re-introduce purpose or direction into nature by the back door. Darwinists often conveniently forget that chance is blind and lapse into using phrases like 'selection-pressure' (a favourite phrase of both Mayr and Dobzhansky), imagining that natural selection can place an order with random mutation like diners choosing from a restaurant menu.

Waddington has even proposed a (hypothetical) mechanism whereby the life of the individual could actually influence mutation of the genes contained in its individuals germ cells, thus influencing the genetic character of its offspring. For committing this heresy of Lamarckism – the most heinous crime in the Darwinian calendar – Waddington was soundly denounced by his fellow evolutionists, especially Jacques Monod. Waddington's suggestion is examined in a later chapter because it is of considerable interest.

An unusually clear example of the corrupted vulgate version of neo-Darwinism in practice occurred with the recent broadcast of an Open University programme on BBC television. The broadcast concerned certain species of wild flower that had adapted to life on railway embankments dug 100 years or so in the past, in rocks containing highly toxic minerals, such as arsenic, antimony and lead. The programme's presenter explained to Open University biology students that here was an example of natural selection and evolution in action. The cuttings had been dug in rocks where no flower could survive, he said, placing an extraordinary environmental demand on nature. Yet, within 100 years, species of wild flower had evolved which were able to tolerate and actually to thrive in the highly toxic conditions where all usual varieties had withered away.

The magnitude of this claim and the magnitude of its falseness is simply breathtaking. First, no specific change has occurred: no new species have come into existence, so the claim that evolution has occurred is simply untrue. Second, the claim that the appearance of plants which are toxin-resistant is an example of natural selection is equally false. What has happened is precisely the same as in the case of so-called 'industrial melanism' in moths: the plants unable to tolerate toxic soils all died, leaving the ground clear for plants which are not poisoned by the metals in question. To imagine that a new species of plant came into existence because workmen dug over the ground is reminiscent of the eighteenth-century belief that maggots in cheese represented the spontaneous generation of life. The presenter also said that the appearance of these 'novel' types was *in response* to the evolutionary demand of natural selection for just such a plant.

This kind of thinking is symptomatic of the confusion that the teaching of neo-Darwinism leads to. Though many lecturers and teachers are sufficiently well informed to know that something is amiss, they quieten their consciences with the reflection that what they are passing on to their students is merely the 'popular' form of the theory, which in its true form remains inviolate and inviolable.

PART FOUR

Creation

CHAPTER 14

Paradigm Lost

In 1962, Thomas Kuhn astonished his academic contemporaries by proposing that scientific theories should be looked on not as dealing only with pure objective facts, but rather as systems of belief relating to a wider context: a frame of reference consisting of interlocking scientific, social and even political ideas. This ideological context, which Kuhn terms a paradigm, is implicitly agreed upon by scientists who subscribe to a particular theory and who share the same world-view.[1]

The power of such a paradigm, says Kuhn, is so great that some scientists will continue to believe it even in the face of contradictory evidence (a phenomenon dubbed cognitive dissonance by psychologist Leon Festinger in 1957). This blinkered dogmatism continues until new evidence is overwhelming and a new theory deposes the old – a 'global paradigm shift' occurs.

Such an ideological context can be found in anthropology in the nineteenth century when most Victorian scientists shared the implicit belief that the coloured races were genetically inferior to the white European race. Because the belief in the genetic inferiority of – for instance – the Australian aborigine was widely shared by scientists, then scientific 'evidence' was brought forward to substantiate this viewpoint and was generally accepted: evidence of the aborigine's Stone-Age level of cultural attainment, his coarse features, supposed low intelligence and brutal behaviour.

Thomas Huxley, who was Darwin's leading supporter, observed that 'No rational man, cognizant of the facts, believes that the Negro is the equal, still less the superior, of the white man.' Darwin himself founded much of his evolutionary thinking on equally racist ideas. He indicated in *The Descent of Man* his belief that the Negro races were more closely related to the apes than white people and also his belief that 'At some future period, not very distant as measured by centuries, the civilised races of man will almost certainly exterminate and replace the savage races throughout the world.'[2]

Today, few scientists would maintain that such beliefs were justifiable on grounds of observation and measurement – not because the evidence has changed, but because the ruling paradigm of anthropological science has changed. Institutionalised racism withered under the twin effects of the decline of imperialism and the rise of civil rights movements.

One consequence of a scientific theory occurring in an ideological context is that much of the evidence which apparently supports that theory actually merely supports its acceptability to scientists and members of the community. Few people in Darwin's day questioned belief in the inferiority of the Australian aborigine, simply because that belief was part and parcel of the world-view of Europeans of the imperial Victorian age. But that implicit belief became in turn the foundation for the *scientific* view that all the races of mankind represented an evolutionary spectrum, ranging from the genetically 'undeveloped' aboriginal type to the genetically 'advanced' white European type – and thus appeared to be evidence in favour of the theory of Darwinian evolution itself.

In much the same manner, Darwin's theory became buttressed around at an early stage with a powerful array of supporting evidence, held to confirm its basic principles, which in fact represented nothing more than the assumptions of the ruling ideology of Darwin's era. These assumptions concerned a whole broad range of natural phenomena which were minutely described, such as the persistence of vestigial organs in the human body, left behind by evolution; and the recapitulation of former evolutionary stages by embryos.

Since the ruling ideology, the paradigm, of the life sciences has changed since Darwin's day, the assumptions and ideas which formerly acted as supporting evidence for his theory have melted away like snow on a spring morning. The large mass of peripheral evidence for the theory has gradually been eroded by further discoveries, more accurate observation, and science's changing world view. As is often the case with Darwinism, however, though these former assumptions have been exposed as without foundation, they somehow linger on in the popular evolution mythology and continue to be referred to in textbooks and lectures.

A case in point is the existence of 'vestigial' organs in the human body: organs deemed by evolutionists to have become redundant through the action of evolution. In his 1895 book *The Structure of Man*, Ernst Wiedersheim lists eighty-six organs of the human body supposed to have lost their function, and to be mere appendages which time and further evolution would no doubt dispel entirely from the human frame.[3] (*Encyclopaedia Britannica* currently gives the total of redundant human organs as 'more than 100'.) The list includes organs such as the pineal gland, the thyroid gland, the thymus, the coccyx, the appendix, the ear muscles and the tonsils.

The claims for vestigial organs have been examined by S. R. Scadding of the Department of Zoology at the University of Guelph, Ontario. Scadding's principal conclusion is that 'On the basis of this analysis, I would suggest that Wiedersheim was largely in error in compiling his long list of vestigial organs. Most of them do have at least a minor function at some point in life.'[4]

An example is the pineal gland, located between the hemispheres of the brain and long believed to be a degenerate eye serving no function. Although still something of a mystery, the pineal body is now known to be an endocrine gland (one that works through the bloodstream) thought to be of importance in triggering growth cycles and sexual development in the young individual. It is currently thought that the pineal gland secretes a hormone called melatonin and that this in turn regulates sexual development.

Similarly, the functions of the thyroid gland and thymus were previously unknown and they were assigned 'vestigial' status, until their true functions were elaborated. Removal of the thymus gland in adults has no effect and so it was considered without function. It was discovered in the 1970s that the thymus makes a vital contribution in early infancy to the development of the body's immune system. The thyroid, too, is now known to be an endocrine gland which secretes two hormones vital to metabolism and growth.

Of two famous examples – the appendix and the coccyx – Scadding says:

Anatomically the appendix shows evidence of a lymphoid function since the submucosa is thickened and almost entirely occupied by lymphatic nodules and lymphocytes. There is experimental evidence as well that the vermiform appendix is a lymphoid organ which acts as a reservoir of antibody producing cells. The coccyx serves as a point of insertion for several muscles and ligaments including the *gluteus maximus*. Similarly, for other 'vestigial organs' there are reasonable grounds for supposing that they are functional albeit in a minor way.

From the point of view of human anatomy studies, it matters little that an organ is believed to be useless but is later discovered to have a useful function. From the point of view of evolution theory, it matters considerably, since the supposed 'vestigial' character of such organs has been adduced as evidence of evolution in action. It remains to be seen how many other human organs which are currently supposed to be vestigial turn out to have equally important functions. In the meantime, it would be unscientific, to say the least, to claim them as vestigial.

Once again, few scientists today would take seriously such arguments. But as usual, the existence of vestigial organs is still referred to in school biology lessons and some textbooks, because it seems 'reasonable'. Simpson, for example, in his book on the evolution of horses, describes the human coccyx as being a vestigial organ, homologous with the ape's tail and with no modern purpose.[5]

Like other branches of science, Darwinism has been led down some seriously wrong turnings over the past century by over-enthusiastic individuals. No errant scientist has been more thoroughly disowned by his colleagues than German zoologist Ernst Haeckel. Haeckel performed a service to zoology by coining the handy term 'ecology'. Unfortunately he also conceived the 'biogenetic law' – the idea that the developing embryo passes through or recapitulates stages in the evolution of its entire phylum (its ancestral tribe or race). An unstoppable creator of neologisms, Haeckel asserted in his 1876 book, *General Morphology of Organisms*, that 'Ontogeny recapitulates Phylogeny'.[6]

What Haeckel (and a substantial number of early followers including Darwin) believed was that the human embryo started life resembling a single-celled marine organism, developed into a worm with a pulsating-tube heart, then into a fish with gill slits and a two-chambered heart, into an amphibian with a three-chambered heart, into a mammal with a four-chambered heart and a tail (for swinging through the trees) and finally into a human baby. These various stages involved the embryo exhibiting vestigial remnants of former evolutionary stages (such as gills) through which it was obliged to pass by natural law in order to reach its new, higher stage of evolution.

The biogenetic law is no longer taken seriously by embryologists, but once again the idea has passed into evolutionary myth and is still to be found in some textbooks as well as being referred to in school and university lectures. Although abandoned as having the status of a scientific law, the feeling persists that there is 'something in it'. The trouble with Haeckel's law is that the observations it seeks to offer as evidence for evolution theory come not from nature but – like the Linnaean system – from a human viewpoint, and rely for their force on purely superficial resemblances. The human embryo is never a single-celled marine organism, nor does it ever live in a marine aquatic environment. It never possesses gill-slits nor does it ever breathe, but takes oxygen from its mother's blood stream directly. The human embryo does develop folds of skin superficially resembling gills but they are not gills. They are structures that become the lower jaw, tongue and other organs of the throat.

Equally, the order in which events occur superficially resembles the order of supposed evolution but, as Dr A. J. White has pointed out, is actually different.[7] It is true that the human embryo begins with a single-chambered heart which develops into two chambers. But this early structure then reverts to a single chamber again before re-developing later to two, three and finally four chambers. No Darwinist has so far suggested that the phylum from which humans descended underwent evolution from a two-chambered heart to a single chamber, since this would be a backward step, and bad news for neo-Darwinists.

Although no professional scientist today would consciously admit to believing in the biogenetic law, even the most eminent Darwinists are prone to slip into the error of betraying their real beliefs when not on their guard. Sir Gavin de Beer was a professional embryologist (as well as Director of the British Museum of Natural History). In his 1964 *Atlas of Evolution,* he is careful to disavow Haeckel's 'law' that Ontogeny recapitulates Phylogeny, yet only a few pages later, in talking about the evolution of the eye, he remarks: 'There can be little doubt that the series of stages . . . through which the eye passes in embryonic development is a repetition of the manner in which it evolved.'[8]

The concept of recapitulation of past evolutionary stages was an important one for evolution theory not only in embryology, since it could be used, and was used, to explain a whole range of common observations from the natural world which contradict the fundamental idea of progressive evolution. If species become progressively better adapted to their environment (by developing eyes with lenses and colour vision for instance, or by developing a shape and colouring which mimics other creatures or other natural objects), then one would expect the fossil record to show such cumulative complexity through time.

In fact this is not what the fossil record shows. Sometimes, the anatomy of creatures becomes more complicated (often in bizarre and apparently senseless ways like the skulls of some dinosaurs or the gigantic antlers of the Irish elk) but they are succeeded in the rocks by remains of creatures who become simpler again – and then complicated again. The solution to this mystery, said Darwinists, was that the simpler creatures were merely degenerate recapitulations of their ancestral forms. As described in an earlier chapter, Hyatt arranged the ammonite sequence of the liassic rocks in a palpably incorrect order, based solely on the concept of recapitulation.

'Recapitulation' was useful to evolutionists in other ways. If it were possible – even in principle – for an organism to exhibit anatomical characteristics of its phylum then perhaps evolution might affect its anatomy in different but related ways? Thus some evolutionists have postulated such phenomena as 'proterogenesis' (the appearance of ancestral features in the young of the species) and 'convergence' (a similar anatomical result being reached by evolution in different species through chance mutation coupled with natural selection).

Some neo-Darwinists have been profoundly disenchanted by this proliferation of so-called evolutionary effects and have even been moved to complain. An obviously exasperated Mayr wrote in 1960, 'The attempt to "explain" genetic and selective processes by all sorts of fancy terms like "pedogenesis", "pal-ingenesis", "proterogenesis", and whatnot have had a stultify-ing effect on the analysis. The less said about this type of literature, the better.'[9]

Most of these false avenues are no more than minor embar-rassments to neo-Darwinists. However, it is the concept of 'convergence' (considerably more important than those com-plained of by Mayr) that spotlights the greatest weakness in the synthetic theory and one which, though raised before, has yet to be satisfactorily addressed – unless it is to become a suitable candidate for Kuhn's global paradigm shift. The weakness has to do with the geological events referred to in an earlier chapter, the break-up of the original super-continent of Pangaea into the present-day land masses, thus separating the plant and animal populations of those continents and – according to neo-Darwi-nists – allowing them to evolve in isolation. Uniformitarians place this event towards the close of the Mesozoic era, that is somewhere in the region of 65 million years ago, according to currently accepted geochronometry.

At the time the present continents were formed, the life they contained was very different from that of today. The dominant life forms were the dinosaurs. The only representatives of the mammals (our own branch of the animal kingdom) then alive were tiny shrew-like creatures. It has been proposed that the reptiles dominated every available ecological niche so effecti-vely that the mammals were hardly able to get a toehold (Harvard's Stephen Jay Gould describes them as living in the nooks and crannies of the reptilian world), and it was only after the mysterious mass extinction of dinosaurs and thousands of other species at the end of the Mesozoic era that mammals were able to begin their rise to dominance, culminating in the appear-ance of humans.

Practically all the mammals that have appeared are either placental (bearing young until fully developed, like humans) or marsupial (giving birth prematurely and nurturing the young in a pouch, like kangaroos). The marsupial mammals are confined to Australia and South America, and are said to have evolved uniquely in those environments, while at the same time placen-tal mammals were evolving elsewhere.

The key factor about the evolution of the marsupials is that a large number of modern marsupial animals exist which – apart from the pouch and child-rearing habits – are identical with placental mammals to an extraordinary degree. This is no mere general similarity of anatomical detail, but an almost perfect duplication of distinctive species like cats, rats, wolves, moles, flying squirrels, ant-eaters and others. In addition there are distinctive marsupials which exist only in Australia, such as the koala and the kangaroo.

How does it come about that in widely separated environments, the same tiny shrew-like ancestral mammal of 65 million years ago should evolve on strictly parallel lines to produce virtually the same range of large mammals today? The Tasmanian marsupial wolf is a virtual carbon copy of the European wolf. The marsupial flying phalanger is practically identical to the placental flying squirrel, as are the marsupial jerboa and the placental jerboa. When the skulls of the two wolves are placed side by side, it would take a very experienced anatomist to tell them apart.

The question for neo-Darwinists is: how can a mouse-like creature have evolved into two identical wolf-like creatures (*and* two identical moles, etc.) on two different continents? Doesn't this coincidence demand not merely highly improbable random mutations, but miraculous ones? According to Simpson in *The Meaning of Evolution* the answer is simple. This convergence comes about through the 'selection of random mutations'.[10] In a different book, the same author concludes that 'Tasmanian and true wolves are both running predators, preying on other animals of about the same size and habits. Adaptive similarity involves similarity also of structure and function. The mechanism of such evolution is natural selection.'[11]

As Arthur Koestler observed of this example, 'One might as well say, with the wisdom of hindsight, that there is only one way of making a wolf, which is to make it look like a wolf.'[12]

The stupendous inadequacy of Simpson's explanation, and the almost casual way in which neo-Darwinists have batted aside the marsupial problem, is, I think, a symptom of their uneasiness over the issue – perhaps a symptom of cognitive dissonance. The response reveals the synthetic theory's inability to explain a key real-life biological problem. But more than this, the existence of identical evolutionary outcomes in isolated environments is the strongest possible indication that random mutation and natural selection are incapable of explaining the origin of species, and at the same time the strongest possible indicator of some other important process or processes at work, acting in some way to limit or direct the repertoire of evolution. An exploration of what those processes may be will have to wait until a later chapter.

In discussing neo-Darwinism as a paradigm, a key question seems to me to be, what is it exactly that drives Darwinism? I call this a key question because virtually every apparent scientific discovery that stimulated the theory in the first place and later assisted with its development has now been found to be false. The variation Darwin observed in the Galapagos was no more than sub-specific variation of the common kind, incapable of leading to novelty; his 'natural selection' is now seen to explain nothing except in the trivial sense that the fit survive and those who survive are fit; the fossil record in the 'areas most likely to afford remains' have been thoroughly searched and palaeontologists have found no fossil evidence linking mankind with extinct apes. There are no fossil 'missing links' in the geological record. Even de Vries's discovery of 'mutations' was a mistake, since the plant he observed turned out to be a freak; Dobzhansky's attempts to significantly alter the genetic structure of *Drosophila* were a failure; all attempts to discover and examine a beneficial mutation have met with failure, the only mutations observed are fatal defects; so-called 'industrial melanism' in moths turns out to be not 'natural selection' but an ordinary population shift; so-called orthogenesis in fossil horses turns out not to be an example of evolution but a group of disconnected species. The Earth is probably not billions of years old and may only be thousands; sedimentary rocks do not take millions of years to form, but hours to lithify. And so on, and on, and on.

If everything that made evolutionists believe in Darwin turns out to be false, why do they go on believing? The answer I think is that, on one hand, there is no credible alternative and, on the other, whatever theory we adopt must seem reasonable. Neo-Darwinism is simply the only 'reasonable' theory we have. Most scientists and teachers in the Earth sciences and life sciences still feel that some form of Darwinian evolution is the only reasonable hypothesis. And they hold this perfectly natural and understandable belief not because of the evidence, but despite the evidence.

CHAPTER 15

Angels Versus Apes

Direct conflict and confrontation with religious belief was built into Darwinism from the outset. Darwin expected trouble, and being a retiring sort did not relish the prospect. But his great champion, Thomas Huxley, certainly did savour the cut and thrust of scientific debate and after urging Darwin to make his findings public, Huxley confided to Darwin on the eve of publication that he was 'sharpening up my claws and beak in readiness'.

In less than a year, Huxley found the major opportunity he sought for public debate at the notorious British Association meeting in Oxford in June 1860. The debate over Darwin's newly published theory took place in the library of the university's museum at the end of a week of meetings where the explosive issue was never far below the surface, threatening to ignite at any time. By Saturday, tension was high and some 700 people crammed noisily into the library, including groups of cheering and counter-cheering students, forcing speakers to shout to make themselves heard above the din.

The combatants this cheerful mob had come to hear were Huxley, championing Darwin, and Samuel Wilberforce, Bishop of Oxford and Fellow of All Souls, a scintillating orator who represented the theological faction. According to Charles Lyell, who was present, Wilberforce began well, launching into a series of calculated and savage attacks which drew loud applause from his supporters. Having warmed up his audience, however, Wilberforce made the tactical error of launching a personal attack on Huxley.

Turning to the young geologist, Wilberforce asked him whether he was related to the apes on his grandfather's side or his grandmother's.

Huxley replied to the bishop's scientific arguments with 'force and eloquence'. He then addressed the personal remark and told Wilberforce:

A man has no reason to be ashamed of having an ape for a grandfather or grandmother. If I had a choice of ancestor, whether it should be an ape, or one who having a scholastic education should use his logic to mislead an untutored public, and should treat not with argument but with ridicule the facts and reasoning adduced in support of a grave and serious philosophical question, I would not hesitate for a moment to prefer the ape.[1]

Huxley was judged by Lyell to have got the better of the debate on this occasion but, then as now, nothing is ever settled merely by debating the question of Creation versus Evolution. The two sides are as deeply entrenched in the 1990s as they were in the 1860s. Attacks by religious fundamentalists remained confined to the debating chamber for some decades after Huxley confronted the Bishop. But in the early twentieth century, evolution began to be taught in schools and this gave religious groups a battleground on which to fight. The first result was the famous Scopes trial in Tennessee in 1925.

In March 1925, Bible fundamentalists in Tennessee instigated the passing by the state legislature of a law forbidding the teaching of any doctrine denying the creation of mankind as taught by the Bible. The American Civil Liberties Union decided to contest this law and a young school teacher, John Scopes of Dayton, volunteered himself as a defendant. The trial became a confrontation not only of fundamentalists versus evolutionists but also of two great American public figures, William Jennings Bryan, who prosecuted, and Clarence Darrow who defended. Bryan was a dedicated bible scholar and champion of many causes although he prosecuted in this case. Darrow was an advocate of freedom of expression and a highly successful criminal lawyer with an appetite for defending a mighty cause.

In the event the trial was a disappointment to both sides because the judge ruled that the issue of Darwinism itself was not to be tried. The trial was to confine itself solely to the question of whether Scopes had broken state law, which was not disputed. Scopes was found guilty and fined $100, although the conviction was later quashed by the supreme court on the technical grounds that the penalty imposed was beyond the power of the court to impose. The law under which he was prosecuted remained on the statute books of Tennessee until 1967.

Although the proceedings of the Scopes trial contributed nothing concrete to the debate, they do provide some further insight into the spread of Darwinist dogma. For some interesting exhibits were introduced in evidence at the trial in order to establish a factual basis for the teaching of evolution. These exhibits included our old friend 'Piltdown Man' together with a tooth which was the sole fossil remains of his American counterpart *Hesperopithecus* – 'Western ape man'.

The Piltdown finds had not yet been recognised as the creation of a practical joker who had cleverly associated an orangutan's jaw with a human skull. The remains of the 'Earliest Englishman' were described to the hushed courtroom as positive proof of mankind's simian ancestry. The other fossil was a tooth that had been found in 1922 in Pliocene deposits in Nebraska, by an amateur geologist, Harold Cook. Cook sent the tooth to Henry Fairfield Osborn, eminent Director of the American Museum of Natural History. Osborn believed he could see anatomical features of both ape and man in this tooth and that it proved man had descended from apes in America as well as Europe and Asia; that the new world could lay claim to a little palaeontological glory just as well as the old.

Some years after the trial, an expedition from the American Museum of Natural History returned to the place where Cook had made his discovery and excavated a number of similar teeth. These showed that *Hesperopithecus* was in fact not a man but an extinct pig.

So, in so far as evidence was produced to provide a factual basis for teaching evolution, that evidence was actually entirely bogus, resulting in one case from a deliberate attempt to deceive (though not that of the defence), and in the other from a piece of overenthusiastic identification by the American Museum of Natural History, keen to keep up with the palaeontological Joneses.

Although the trial itself accomplished little, the nature of fundamentalist objection gradually changed later in the century and became far more sophisticated. In 1964, for instance, religious leaders in Texas objected to the State Board of Education approving biology textbooks containing Darwin's theory. On this occasion, the objectors were over-ruled but in 1969 similar objections were successful – in California of all places – where the State Board of Education decided that textbooks in future should present Darwinism as merely one of many competing theories. In support of their case, the protesters quoted Mayr as saying, 'The basic theory is in many instances hardly more than a postulate and its application raises numerous questions in almost every concrete case.'[2]

The greater sophistication of religious objectors consists in their arguing that neo-Darwinism is merely one theory among many and must not be taught as the sole repository of truth. A further level of sophistication has come about in the past thirty years with the rise of creationists, who believe that God is the prime cause of the universe and the creator as described in the Bible, but some of whom believe that it is not necessary to insist on a literal interpretation of Genesis. Some creationists believe, for example, that the six days taken to create the Earth may be construed as a metaphor. One particular variant of this group is scientific creationism, many of whose adherents are Christians who are also professional scientists. This group has proved quite a thorn in the side of proponents of the synthetic theory, because their expertise has been employed to turn the tables on their opponents by applying their own methods.

The strategy of this group has been not to emphasise the extent to which Darwinism contradicts the religious teaching of the Bible, but the extent to which it is contradicted by other scientific evidence. Like neo-Darwinists, creationists have had their ups and down. For some years they believed they had found fossil footprints of humans contemporary with those of dinosaurs, at the Paluxy River in Texas. However, more detailed examination in 1970 by Neufeld, Brand and Chadwick found that the tracks had been made by a three-toed bipedal dinosaur whose footprint happened to look rather human.[3] On the whole, though, many of the purely scientific arguments advanced by creationists have been eminently sound and merit serious consideration, though they rarely if ever receive such attention.

But if religious fundamentalists have been slowly trans-formed (I am tempted to say 'have evolved') into something closely resembling scientists, then evolutionists have also been metamorphosed, ironically into something closely resembling a religious order. Princeton's professor of biology, Edwin Conklin, observed in 1943, 'The concept of organic evolution is very highly prized by biologists, for many of whom it is an object of genuinely religious devotion, because they regard it as a supreme integrative principle. This is probably the reason why severe methodological criticism employed in other departments of biology has not yet been brought to bear on evolutionary speculation.'[4] This insight by Conklin is exquisitely apt. He equates the fervour of fundamentalists with that of Darwinists, in particular the capacity of both groups to speculate endlessly while avoiding contact with the hard truths of nature. When one searches the literature of evolution for solid ground and indisputable fact, one finds instead only shifting sands. The only bedrock to be found is that of faith: faith in God; faith in mutation and natural selection. Ultimately there is little to choose between them.

Some Christians have adopted a compromise philosophical stance; accepting synthetic evolution as inescapable scientific fact, but continuing to believe in a creator as the first cause of the world in which evolution takes place: that evolution is the mechanism of divine creation. There is no logical or scientific objection in principle to such a belief, any more than there is a logical or scientific objection to belief in special creation. In both cases (I should say 'in all three cases' and include neo-Darwinism) there is no foundation of evidence or experiment on which to base a scientific case either for or against. Thus ultimately it becomes a matter of belief or faith.

Although the debate remains unchanged since Darwin's time, and even the rapid strides made by scientific research have failed to contribute decisively to the arguments one way or the other, a new dimension has entered the debate in recent years which has *appeared* to offer some decisive evidence. This new dimension is the advent of artificial intelligence in the form of the computer. The information-handling capability of electronic data processing, with its obvious analogy to DNA, has been gleefully enlisted by computer-literate Darwinists as offering powerful evidence for the synthetic theory; while their opponents have often reacted to this new found high-tech confidence with a kind of sheepish bewilderment.

There are a number of important applications based on computer simulations that certainly deserve to be examined seriously. A good example of this is in the field of aircraft wing design where computers have been used by aircraft engineers to evolve the optimum aerofoil profile. In the past, wing design has been based largely on reiterated trial and error methods. A hypothetical wing shape is drawn up; a physical model is made and is tested aerodynamically in the wind tunnel. Often the results of such an empirical design approach are predictable: in general lengthening the upper wing curve in relation to the lower increases the upward thrust obtained. But sometimes results are very unpredictable, as when complex patterns of turbulence combine at the trailing edge to produce drag, which lowers wing efficiency, and even causes destructive vibration.

Engineers at Boeing Aircraft tried a new approach. They created a computer model which was able to 'mutate' a plain wing shape at random – to stretch it here or shrink it there. They also fed into the model the rules that would enable the computer to simulate testing the resulting design in a computerised version of the 'wind tunnel' – the rules of aerodynamics. This process has resulted, say the engineers, in obtaining wing designs offering maximum thrust and minimum drag and turbulence, more quickly than before and without any human intervention once the process has been set in motion.

The engineers have gained enormous savings in time against previous methods and the success of the computer in this field has given rise to a new breed of application dubbed 'genetic software'. Indeed, on the face of it, the system is acting in a Darwinian manner. The computer (an inanimate object) has produced an original and intelligent design (comparable, say, to a natural structure such as a bird's wing) by random mutation of shape combined with selection according to rules that come from the natural world itself – the laws of aerodynamics. If the computer can do this in the laboratory in a few hours or days, what could nature not achieve in millions of years?

The fallacies on which this case is constructed are not very profound but they do need to be nailed down. In a recently published popular primer on microbiology, Andrew Scott's *Vital Principles*, this very example is given under the heading 'The Creativity of Evolution' and the process itself is called 'computer generated evolution' as though it were analogous to an established natural process of mutation and selection.[5]

The most important fallacy in this argument is the idea that somehow a result has occurred which is independent of, or in some way beyond, the engineers, who merely set the machine going by pressing a button. Of course, the fact is that a human agency has designed and built the computer and programmed it to perform the task in question. This begs the only important question in evolution theory: could complex structures have arisen spontaneously by random natural processes *without any precursor*? Like all other computer simulation experiments, this one actually makes a case for special creation because it specifically requires a creator to build the computer and think up and implement the programme in the first place.

However, there are other important fallacies too. The only reason that the Boeing engineers are able to take the design produced on paper by their computer and translate that design into an aircraft that flies is because they are employing an immense body of knowledge – not possessed by the computer – regarding the properties of materials from which the aircraft will be made and the manufacturing processes that will be used to make it. The computer's wing is merely an outline on paper, an idea: it is of no more significance to aviation than a wave outline on the beach or a wind outline in the desert. The real wing has really to fly in the air with real passengers and the decisive events that make the idea into a reality are a long and complex sequence of human operations and judgements that involve not merely the shaping and fastening of metal for wings but also the design and manufacture of airframes and jet engines. These additional complexities are beyond the capacity of the computer not merely in practice but in principle, because computers cannot make even a cup of coffee, let alone an airliner, without being instructed every step of the way.

In order for a physical structure like an aircraft wing to evolve by spontaneous random means, it is necessary for natural selection to do far more than select an optimum shape. It must also select the correct materials, the correct manufacturing methods (to avoid failure in service) and the correct method of integrating the new structure into its host creature. These operations involve genetic engineering principles which at present are quite unknown. And because they are unknown by us, they cannot be programmed into a computer.

There is also an important practical reason why the computer simulation is not relevant to synthetic evolution. That is because an aircraft wing differs from a natural wing in a fundamental way. The aircraft wing is passive, since the forward movement of the aircraft is derived from an engine. A natural wing like a bird's, however, has to provide upthrust and also the forward motion necessary to generate that lift and is thus a complex articulated active mechanism. The engineering design problem of evolving a passive wing is merely a repetitive mechanical task – that is why it is suitable for computerisation. No one has so far suggested programming a computer to design a bird's wing by random mutation because the suggestion would be seen to be ludicrous. Even if all of the world's computers were harnessed together, they would be unable to take even the first elementary steps in designing a bird's wing, unless they were told in advance what they were aiming at and how to get there.

If computers are no use to evolutionists as models of the hypothetical selection process, they are proving invaluable in another area of biology, one that seems to hold out much promise to neo-Darwinists – the field of genetics. Since Watson and Crick elucidated the structure of the DNA molecule, and geneticists began unravelling the meaning of the genetic code, the centre of gravity of evolution theory has gradually shifted away from the earth sciences – geology and palaeontology – towards microbiology.

This shift in emphasis has occurred not only because of the attraction of microbiology as holding the answers to many puzzling questions, but also because the traditional sciences had proved ultimately sterile as a source of decisive evidence. The gaps in the fossil record, the incompleteness of the geological strata, the ambiguity of the evidence from comparative anatomy, ultimately caused Darwinists to give up and try somewhere else. Thanks to microbiology (and computer science), they now have somewhere else to try.

In the microscopic version of the synthetic theory, Darwinism is said to have been elegantly confirmed by discoveries at the cellular level. Patient research has elucidated the main nucleotide sequences of the DNA of humans and of other animals and plants. When compared, say microbiologists, these DNA sequences demonstrate the extent to which all the organisms are related.

There are two particular areas in which evidence from

microbiology is used to substantiate neo-Darwinism. The first is the field of blood serum. If human blood is injected into a rabbit, the rabbit produces anti-human serum which, when mixed with human blood, causes clumping of 100 per cent of the blood protein. This is taken to be a measure of extremely high chemical affinity. The anti-human serum precipitates blood of other species as follows: gorilla, 64 per cent; orang-utan, 42 per cent; baboon, 29 per cent; ox, 10 per cent; horse, 2 per cent and kangaroo, 0 per cent. These findings are also said to agree with results from comparative anatomy, embryology and palaeontology.

The second area is that of studies of the protein enzyme called cytochrome c, which has the same general chemical structure in bacteria, fungi, plants and animals, but with differences in its detailed structure in each case. Darwinists believe that these similarities and differences mirror the tree of evolutionary descent of all organisms and show how closely or distantly they are related.

But is this really what the microscope shows? Isn't it simply the case that these studies show at the microscopic level the similarities in nature that comparative anatomists have *already* noted at the macroscopic level? True, the DNA sequences explain the basis of the macroscopic structure and are thus a very important step forward for biology. But the important point in the context of evolution theory is that they do not, as Darwinists claim, explain how those structures came into being.

The fact that all mammals have four limbs is (presumably) a reflection of a basic nucleotide sequence which determines this anatomical characteristic, rather than, say, three limbs or five. The fact that the four-limbed genetic code is found in a wide variety of animals which are presumed to be more or less closely related to humans is a discovery of immense interest and significance. However, the discovery does not necessarily entail the conclusion that all those organisms share a common ancestor.

Consider a future archaeologist from Mars picking through the debris of our civilisation. He may find the remains of a large number of cars; he may determine by measurement and analysis that they have many very close similarities of design and construction, even down to the microscopic composition of the steel and the pitch of threads on the bolts. He may be inclined to believe that they are all related in the sense that they are all based on the same design (share a common 'ancestor') when this is not in fact the case. What has happened is that they are independently derived by individual manufacturers, but using common design principles and common technological processes, and drawing their materials from a common pool.

If it is a property of living organisms that a unique sequence of nucleotides generally leads to the formation of four limbs, then whenever you encounter a four-limbed creature, you would expect to find that particular sequence of nucleotides – regardless of how that creature has evolved or from what ancestors it has descended, in just the same way that wherever you encounter flight, you expect to find a wing. The possession of a wing is not diagnostic of common ancestry, for not all winged creatures are related. In fact Darwinists believe that wings have evolved separately a number of times (for example in birds, bats and flying reptiles). But presumably the DNA sequences for those wings will have at least some instructions in common. And while common macroscopic features like this are not taken as primary evidence of genetic relationship, for some reason microscopic structure exercises a fascination for modern Darwinists that blinds them to it signifying anything less than a blood relationship.

But in any case, even if studies of the similarity of cytochrome c should provide strong circumstantial evidence that all animals are genetically related, or even strong circumstantial evidence that animals have evolved from a common ancestor, what they do not prove is that the mechanism of that evolution was genetic mutation coupled with natural selection – the synthetic or neo-Darwinian model. In fact they take the problem of how animals evolved no further than the original macroscopic equivalent observations – that all mammals share a similar anatomical ground plan.

Darwinist microbiologists seem to have drawn immense comfort from their recent discoveries at the cellular level, behaving and speaking as though the new discoveries of microbiology represent a triumphant vindication of their long-held beliefs over the irrational ideas of vitalists. Yet the gulf between what Darwinists claim for microbiological discoveries and what those discoveries actually show is only too apparent to any objective evaluation.

Consider these remarks by Francis Crick, justly famous as one of the microbiologists who cracked the genetic code, and equally well known as an ardent supporter of Darwinist evolution. In his 1966 book *Of Molecules and Men*, in which he set out to criticise vitalism, Crick asked which of the various microbiological processes are likely to be the seat of the 'vital principle'.[6]

'It can hardly be the action of the enzymes,' he says, 'because we can easily make this happen in a test tube. Moreover most enzymes act on rather simple organic molecules which we can easily synthesise.'

There is one slight difficulty but Crick easily deals with it: 'It is true that at the moment nobody has synthesised an actual enzyme chemically, but we can see no difficulty in doing this in principle, and in fact I would predict quite confidently that it will be done within the next five or ten years.'

He adds the rider that 'we could only synthesise a good enzyme at this stage in our knowledge by precisely imitating what nature has produced over the course of evolution rather than by designing one ourselves from first principles.'

A little later, Crick says of mitochondria (important objects in the cell that also contain DNA):

> It may be some time before we could easily synthesise such an object, but eventually we feel that there should be no gross difficulty in putting a mitochondrion together from its component parts.
>
> This reservation aside, it looks as if any system of enzymes could be made to act without invoking any special principles, or without involving material that we could not synthesise in the laboratory.

There is no question that Crick and Watson's decoding of the DNA molecule is a brilliant achievement and one of the high points of twentieth-century science. But this success seems to me to have led many scientists to expect too much as a result.

Crick's early confidence that an enzyme would be produced synthetically within five or ten years has not been borne out and biologists are further than ever from achieving such a synthesis. Indeed, reading and re-reading the words above with the benefit of hindsight I cannot help interpreting them as saying, 'We are unable to synthesise any significant part of a cell at present, but this reservation aside, we are able to synthesise any part of the cell.'

Certainly great strides have been made. William Shrive, writing in the 1982 *McGraw Hill Encyclopedia of Science and Technology*, says, 'The complete amino acid sequence of several enzymes has been determined by chemical methods. By X-ray crystallographic methods it has even been possible to deduce the exact three-dimensional molecular structure of a few enzymes.'[7] But despite these advances no one has so far synthesised anything remotely as complex as a protein molecule or enzyme.

Such a synthesis was impossible when Crick wrote in 1966 and remains impossible today. Probably it is because there is a world of difference between having a neat table that shows the genetic code for all twenty amino acids (alanine = GCA, proline = CCA and so on), and knowing how to manufacture a protein. These complex molecules do not simply assemble themselves from a mixture of ingredients like a cup of tea. Something else is needed. What the something else is remains conjectural. If it is chemical it has not been discovered; if it is a process it is an unknown process; if it is a 'vital principle' it has not yet been recognised. Whatever the something is, it is plainly impossible at present to build a case either for Darwinism or against vitalism out of what we have learned of the cell.

It is easy to see why evolutionists should be so excited about cellular discoveries because the mechanisms they have found appear to be very simple. But however simple they may seem, no one has yet succeeded in synthesising any significant original structure from raw materials. We know the code for the building bricks; we don't know the instructions for building a house with them.

Indeed, the discoveries of microbiology have raised some rather awkward questions for Darwinists, which they have yet to address satisfactorily. For example, the existence of genetically very simple biological entities, such as viruses, on the face of it supports Darwinist ideas about the origin of life. One can imagine all sorts of primitive life forms and organisms coming into existence in the primeval ocean and it seems only natural that one should find entities that are part way between the living and non-living – stepping stones to life as it were. It is only to be expected, says Richard Dawkins, that the simplest form of self-replicating object would merely be that part of the

DNA program which says only 'copy me', which is essentially what a virus is.

The problem here is that viruses lack the ability to replicate unless they inhabit a host cell – a fully functioning cell with its own genetic replication mechanisms. So the first virus must have come *after* the first cell, not before in a satisfyingly Darwinian progression.

But although the modern shift in emphasis from the macroscopic to the microscopic has failed to shed any new light on evolution, moving the microscope away from fossils to chromosomes did have one unexpected side effect – that of focusing evolutionary thinking back in the very remote past, away from macroscopic structures to the ancestral microorganisms that inhabited the archaic oceans. The unpleasant experiences of Piltdown Man and *Hesperopithecus* gave Darwinists every reason to shun the relatively recent historical past and mankind's own ancestry in favour of the more exciting and more promising world of the microscope. Microbes replaced men as the focus of evolutionists' attention.

Thus it came about that the study of potentially the most important subject of all – our own species – has languished in recent decades, and almost fallen into disrepute. In contrast to the first decades of the twentieth century when human anthropology and the study of human evolution was the most important subject of scientific study, it has today been relegated to virtual obscurity and become the province of just a few talented individuals, working in isolation.

The evolution of the human species has become almost a taboo subject, too hot to handle politically; and equally dangerous scientifically. Riddled with doubt and smarting from numerous embarrassing mistakes, evolutionists have quit the field almost entirely, and apart from heroic individual efforts like the Leakeys and Johanson working in Africa, there have been no significant palaeontological discoveries relating to humans since the Second World War.

You might imagine that the effect of this disillusion and abandonment would be reflected in schools, universities and museums, with reduced interest in human evolution and a dearth of teaching materials. Yet the old myths are more active than ever and reconstructions of our ape-like brutish ancestors and their primitive lives form part of schoolboy folklore in the

1990s, just as they did in the 1930s. Even in major works of science fiction, like Arthur C. Clarke's popular *2001*, or Pierre Boule's *Planet of the Apes*, it is taken for granted that our ancestors were ape-like. And the writers and illustrators of popular works of historical geology spend much time and effort searching for minute 'accuracy' in their reconstruction of ape-like palaeolithic hunters and their environment.

Strangely too, this modern confidence and apparent precision in reconstruction is based not on further discoveries of fact, but takes place *despite* the real discoveries of recent decades – that the evidence for mankind's own evolution is actually non-existent.

CHAPTER 16

Down from the Trees

Less than a decade after the publication of *The Origin of Species*, in 1868, Ernst Haeckel published a monumental work, primarily inspired by Darwin and grandly entitled *The History of Creation*, in which he indulged to the full his predilection for coining new words. This time, Haeckel became even more ambitious and coined not merely a new scientific term but a new generic and specific name for a living creature, but a creature no one had ever seen and for which there was no evidence at all – *Pithecanthropus alalus*, the 'speechless ape-man'.[1] *Pithecanthropus*, said Haeckel, was the link between humans and our ape-like ancestors. When his or her fossil remains were discovered, they would be found to have some ape-like characteristics and some human characteristics. Haeckel was even able to describe some of these characteristics: long arms, short legs with knock knees, a half-erect walk, and a long skull with slanting teeth. And the great biologist was also able to suggest the area of the world where the remains were most likely to be found: the hypothetical ancient continent of Lemuria, stretching from Madagascar to India and across the Indian Ocean to Indonesia. However, added Haeckel gloomily, it is 'ridiculous to expect palaeontology to furnish an unbroken series of positive data'.

In the event, Haeckel's pessimism on this point proved unjustified. For within decades of his prediction, *Pithecanthropus* was found, possessing just the characteristics Haeckel had predicted and in the very spot he had foretold.

An ambitious and talented Dutch anatomist, Eugene Dubois, set sail with his wife and young children in 1887 for the Dutch colony of Java in the East Indies. Dubois had signed up for a spell of service with the Dutch army medical corps. A teacher of anatomy, Dubois was thoroughly versed in the works of

Haeckel and as an experienced geologist and palaeontologist, he was perfectly equipped for the task on which he embarked. The young man was setting out with the avowed intention of being the first to discover concrete evidence of the missing link between human and ape; the first to discover *Pithecanthropus*.

Within two years of arriving in Sumatra, Dubois had persuaded the government to allow him to carry out a complete palaeontological survey of Java under his full-time supervision. He was given convict workers to carry out excavations and military personnel to supervise the digging. Dubois himself, however, did not participate in the field work and contented himself with examining each season's finds on the verandah of his house where they were periodically delivered by the convict crew.

In 1891 Dubois made two important finds amongst the bones dumped on his verandah. The fossils were a tooth and a skull cap which had been found a month apart in the same fossiliferous bed but in locations that were not known exactly because no one was recording the finds. At first, Dubois identified them as belonging to a chimpanzee. Some months later, however, the convicts found a fossil thigh bone in the same bed, a thigh bone which belonged unmistakably to an upright walking human. Dubois now revised his earlier identification and put the femur together with the skull cap and tooth to produce *Pithecanthropus erectus* – 'upright ape-man' – a vindication of Haeckel and the first solid evidence of a missing link. Great efforts were made to secure further finds. Some 10,000 cubic metres of sediments were dug and sieved, but the only additional discovery was another tooth.

The third International Congress of Zoology at Leiden in 1895 greeted the fossils with unanimous recognition as to their importance, but with a mixed reception regarding their interpretation. The president, Rudolph Virchow (founder of the modern science of pathology), cast doubt on the remains belonging to a single individual. Some members felt they were more ape than human, while others felt they were entirely human. A few agreed with Dubois that he had bagged a missing link. Haeckel, who was present, was delighted to have been proved right but was rather circumspect about the finds: 'Unfortunately, the fossil remains of the creature are very scanty: the skullcap, a femur, and two teeth. It is obviously impossible to form from these scanty remains a complete and satisfactory reconstruction of this remarkable Pliocene Primate.'[2]

If he was publicly cautious, though, Haeckel was privately convinced because he paid from his own pocket for a life-size reconstruction to be built which stands today in the basement of the Leiden natural history museum. In common with all reconstructions of ape-men, both three dimensional and pictorial, which have been essayed since Haeckel and Dubois's day, the Leiden statue bears a human-like body and a rather dim, ape-like face. He is gazing with a puzzled frown at a crude knife-like tool clutched in his primitive hand, as though trying to remember with his small brain how he came to be relegated to the museum basement from the fashionable salons upstairs.

Though his statue has been edged away from the public gaze both Haeckel's *Pithecanthropus* and the creature's familiar epithet 'Java Man' still figure prominently in evolutionary mythology, a testament to the staying power of a good story, whatever its true merits.

The story of Dubois's discovery of Java Man, like Gideon Mantell's discovery of the first dinosaur, is a parable of primate palaeontology in the past 100 years. Discoveries are few and fortuitous, yet it is extraordinary how they are always deliberately sought by their discoverers. The reconstructions, the names bestowed and the attributions to human or ape inheritance blow this way and that in the wind of scientific opinion. In the end each find has its supporters and detractors but settles nothing.

This question of attribution has bedevilled every 'missing link' discovery of the twentieth century. The pattern is a recurring one. The remains themselves are always meagre. The first attribution is always that the being whose remains have been discovered shows both human characteristics and ape characteristics, and is therefore a genuine transitional type – a real missing link. Then the attribution is questioned: the characters ascribed to apes are actually within the range of human characters; or ape remains post-date the finds by a large margin; or the reconstruction work is over-imaginative; sometimes simple mistakes of identification are made perhaps due to disease or malformation of bones.

The position today is that all the fossil remains which were previously assigned some intermediate status between apes and humans have later been re-assigned definitely into the

categories of either extinct ape or human, and this re-assign-
ment has been accepted by all but the most fanatical devotees of
this or that fossil.

Strangely enough, although evolutionists from Darwin
onwards have frequently harped on about how unfair it is that
vertebrate remains are very rare and their discovery a matter of
chance, the world's natural history museums are today bulging
with vertebrate remains from Europe, Asia and Africa. Yet as
with all other branches of the animal kingdom, the gaps remain
where there should be transitional species. In the case of
humans, there is not just one gap to be filled (between apes and
ourselves), but many gaps.

First there is the gap between mammals and the rest of the
animal kingdom. So far there is no candidate for the ancestor of
all the mammals except a hypothetical one. No fossil remains
have been found. As A. J. White points out, there are a number
of distinctive anatomical differences between reptiles and mammals, chiefly the articulation of the jaw (mammals have a single
lower jawbone while reptiles have six) and the mechanism of
the ear (mammals have three ear bones, reptiles have one). So
recognising a transitional skeleton ought to be straightforward
if, as evolutionists claim, mammals evolved from reptiles.[3] The
earliest mammals are small rodent-like animals, but there is no
evidence in the fossil record for the evolution of rodents or any
other group of mammals. According to Alfred Romer of the
University of Chicago, in his 1966 textbook *Vertebrate Palaeontology*:

> The origin of the rodents is obscure. When they first appear, in
> the late Palaeocene, in the genus *Paramys*, we are already dealing
> with a typical if rather primitive true rodent with the definitive
> ordinal characters well developed. Presumably of course they
> had arisen from some basal, insectivorous, placental stock, but
> no transitional forms are known.[4]

The same is true of all other mammals from bears to whales and
from walruses to carpenters.

Next there is the gap between the primates and the rest of the
mammals. Again, the candidate for this honour is a hypothetical
insectivore but no remains of this ant-eating ancestor have ever
been found. According to A. J. Kelso in his 1974 *Physical Anthropology*, 'The transition from insectivore to primate is not docu-

mented by fossils. The basis of knowledge about the transition is by inference from living forms.'[5] Unfortunately, Kelso neglects to mention from which living forms this knowledge is inferred or how it is inferred. Perhaps the insectivorous ancestry is inferred from a long nose.

Then there is the crucial gap: the gap between the hypothetical ape-like primate ancestor and ourselves. Despite scores of candidates, the glass cabinet marked 'missing link' remains tantalisingly empty. No primate palaeontologist has gone on record as admitting such a heretical thought but it is hard to resist the conclusion that it is now likely to remain empty. Richard Leakey has quoted fellow palaeontologist David Pilbeam as saying:

> If you brought in a smart scientist from another discipline and showed him the meagre evidence we've got he'd surely say, 'forget it; there isn't enough to go on.' Neither David nor others involved in the search for mankind can take this advice, of course, but we remain fully aware of the dangers of drawing conclusions from evidence that is so incomplete.[6]

To illustrate the dangers of drawing such conclusions, here is a summary of the stories of just a few fossils which failed to make it into the glass case, and which, like Java Man, have been relegated to the basement of history.

Probably the most celebrated supposed ancestor of modern humans is the unfortunate gentleman whose remains were discovered by quarrymen in a gravel pit in the Neander Valley, near Dusseldorf, in 1875. The skull cap and limb bones of 'Neanderthal Man' were launched on the world by Hermann Schaffhausen, professor of anatomy at Bonn University, at the same time putting into the language a new synonym for coarse, unintelligent brutality.

Neanderthal man was depicted as a shambling brute, who walked with an ape-like gait, on the edge of his feet, his low sloping brow denoting his retarded mentality and anti-social tendencies. He was unquestionably, thought Schaffhausen, part-ape, part-human and ancestral to modern humans.

It was not until the 1950s, by which time many similar remains had been found in Europe, Africa and Asia, that Neanderthal man was seriously re-evaluated. It was found that some of the original type material belonged to an individual whose bones were thickened and deformed by osteo-arthritis and that Neanderthal man's posture was probably the same as modern humans'. Evidence was also found that far from pre-dating Cro-Magnon (modern) humans, the Neanderthals lived at the same time and possibly mixed freely with Cro-Magnons.

Neanderthals sewed clothes from animal skins, used fire for cooking, built shelters and gave their dead a ritual interment which included placing flowers in the grave. Finally it was observed by Cave and Strauss writing in *The Quarterly Review of Biology* that if he were given a bath, a collar and tie, he would pass unnoticed in the New York subway.[7] Today Neanderthal man is classed as a member of the species *Homo sapiens* and any of us could be among his descendants.

Raymond Dart was a young Australian anatomist who was appointed professor at Witwatersrand University, Johannesburg, in 1922. Dart's speciality was the evolution of the brain and nervous system and he had worked under Grafton Elliott Smith at University College London on developing the technique of making endocranial casts (casts of the inside of skulls) to get an indication of the development of the brain within. This skull skill was coincidentally to play a prominent part in his discovery.

In South Africa, Dart arranged for workers at the nearby Taung quarry to send him any fossils they found in the limestone rocks being quarried for building stone. In one batch of fossils shipped to his office, Dart immediately recognised a natural endocranial cast made by limestone filling a fossil skull, together with some fragments of the skull itself. It has been remarked before that this endocranial cast had fallen into the hands of one of the three or four people in the entire world capable of recognising its significance.

Dart believed the cast to show distinctly hominid features in the brain structure, far in advance of any living ape, yet still small and underdeveloped for a human. Dart felt that 'by the sheerest good luck', he had been given 'the opportunity to study what would probably be the ultimate answer in the study of the evolution of man'.[8]

He wrote a paper for *Nature* and christened his discovery *Australopithecus africanus* – Southern Ape-Man. (Dart was here following the precedent of the American Museum of Natural History who, as we saw in Chapter 15, christened a pig's tooth

Hesperopithecus – Western Ape-Man.) Ironically, Dart's discovery was scorned by his scientific contemporaries, partly because of his irreverent Aussie style, but mainly because his identification rested on specialised knowledge that only he and two or three others possessed on endocranial casts. Dart's discovery was taken up and championed later by Robert Broom, who discovered many more *Australopithecus* remains, which Broom believed showed among other things that the creature walked upright. Today *Australopithecus* is the only fossil find which stands any chance at all of being placed in the missing link category and is enthusiastically described by Darwinists as ancestral to humans.

The real status of *Australopithecus* as an extinct ape was established as long ago as 1954 by the comparative anatomy research of zoologist Sir Solly Zuckermann and his colleagues. Sir Solly compared in detail three diagnostic characteristics in the bones and teeth of *Australopithecus,* in modern humans and in various apes including the gorilla, chimpanzee, orang-utan, gibbon, and others. The key characteristics are the size of brain; the jaws and teeth; and the posture of the head. By measuring the skulls and teeth of a large number of apes, of fossil australopithecines, and various human specimens, Zuckerman found that *Australopithecus's* head was balanced like an ape, not a human; its brain was the same size as modern apes such as the gorilla; and that the jaws and teeth are predominantly ape-like. According to Sir Solly:

> In the first place, our safest inference from the available facts is that the brains of the fossil *Australopithecinae* did not differ in size or conformation from those of such modern apes as the gorilla. In the second we may conclude that the fossils provide no significant evidence of the major decrease in size of jaws and teeth which is presupposed by the thesis that the *Hominidae* evolved from non-human primate forms. And thirdly the evidence is also clear that the skull of the *Australopithecinae* was balanced on the vertebral column as in apes rather than as in man.

Sir Solly's conclusion is that:

> The safest overall inference that can be drawn from the facts which have been discussed here is that the *Australopithecinae* were predominantly ape-like, and not man-like creatures.[9]

When *Australopithecus* disappeared from the headlines, it was not because of Sir Solly's scientific findings but because of excitement over more missing links – this time from East Africa, where Louis Leakey, his wife Mary and son Richard have made many discoveries in the region around Olduvai Gorge. The principal cause of the excitement was that the Leakeys' discoveries were made in volcanic deposits which unlike the sedimentary limestones of South Africa could be dated by the newly developed potassium-argon method. And using this method yielded a date for the Olduvai Gorge finds of no less than 1.75 million years. This was news indeed. At just the same time, in 1959, Mary Leakey found at Olduvai an almost complete skull which her husband announced to the world as *Zinjanthropus* – East African Man. Notice that the terminology has changed. Dart's terminal 'pithecus' (Ape-Man) has been quietly dropped in favour of the more streamlined 'anthropus' (Man).

Part of the reason for the change of name was Leakey's insistence that his discovery was entirely novel, was not related to Dart's australopithecine discoveries in the south, and was definitely hominid, not an ape. Alas, *Zinjanthropus*, too, fell victim to the curse of all missing links. In 1965 Professor Philip Tobias of Witwatersrand University examined, measured and described the Olduvai fossil skull in the official monograph in which he re-assigned the specimen as *Australopithecus* (*Zinjanthropus*). The Olduvai find was merely a variety of Dart's fossil and was, after all, an ape, worthy only of a mention in brackets.[10]

As far as the age of the fossil is concerned, we saw in Chapter 4 that the potassium-argon method of dating has yielded dates ranging from 160 million to 3,000 million years for rocks formed in a volcanic eruption only 190 years ago. The method is subject to so many separate sources of major inaccuracy that little confidence can be attached to dates stretching back millions of years 'derived' from the technique. If the dating of the associated rock formation is subject to error then we are left in the dark regarding the age of its fossil contents.

One further puzzle remained regarding the area in which the Leakeys had found *Zinjanthropus* and that was the presence of stone tools. If all the fossils found so far were apes, who had made the thousands of tools which littered the Olduvai Gorge? The answer was found in 1964, when once again the pages of *National Geographic*, *Nature* and *The Times* resounded to the discovery of yet another missing link. This time it was a new species of human, man the toolmaker, *Homo habilis* – DIY Man. Again it was the Leakeys who made the discovery.

On this occasion, the find was sparse indeed, consisting only of a lower jaw with teeth, a collarbone, a finger bone and some small fragments of skull. For the first time, a new human species was to be described on the basis of teeth and fragments alone, and in circumstances where the association of the bones as those of a single individual was conjectural – a situation very reminiscent of Dubois and Java Man.

Since 1964 *Homo habilis* has been re-evaluated and it has been suggested that one of the hand bones is a piece of vertebra, that two more could have belonged to a tree-dwelling monkey, and that six came from some unspecified non-hominid. But whatever the merits of the original description, the fact remains that handy man is a human – not a missing link. *Homo habilis* is calculated to have had a small brain: perhaps only half the size of the average modern human's. But, as A. J. White has pointed out, the habilines were also small in stature, so their brain was not small in relation to their body size, rather like modern pygmies.[11]

Other workers have continued to unearth early remains in Africa, notably Donald Johanson and his team working in the Afar region of northern Ethiopia. Johanson has discovered bones and teeth which represent up to 65 individuals, including the famous 'Lucy' which consists of almost half a complete skeleton. Once again, though, the finds have been referred to either *Australopithecus* and hence are apes, or *Homo* and hence are human.[12]

Despite more than a century of energetic excavation and intense debate the glass case reserved for mankind's hypothetical ancestor remains empty. The missing link is still missing.

CHAPTER 17

Hopeful Monsters and Other Optimists

French biologist Jean-Baptiste de Lamarck suggested in his *Philosophie Zoologique* in 1809 that changes in environment would alter an animal's needs, that this in turn would change its behaviour, and that the changed pattern of behaviour would alter its physical structure. As an example, Lamarck pointed to wading birds and suggested that, 'wishing to avoid immersing its body in the water, the bird acquires the habit of elongating and stretching its legs.' Not only did Lamarck think that using organs made them grow (like exercising muscles), he also thought that not using them, made them disappear, like the eyes of the mole.[1]

The problem with this suggestion is that if it were true, then the weightlifter's son would be born with big muscles and the ballerina's daughter would be born dancing. In fact, say Darwinists, characteristics are inherited according to Mendel's law of inheritance: dominant genes preponderate in the offspring, not acquired characteristics.

Today Lamarck is scorned as belonging to the pre-scientific (that is to say, pre-Darwinian) age and the charge of 'Lamarckism' is the most dreaded heresy in the evolutionists' canon. It is strange that such a distinguished biologist should be treated thus, especially as Darwin himself continually flirted with the inheritance of acquired characters and cited many examples, including the case of a man who was reported to have lost his fingers and later produced sons also without fingers. As far as neo-Darwinists are concerned, the matter is settled and the debate is closed. Any backsliding from the straight and narrow line of mutation with natural selection is written off as 'Lamarckism'. But like so many issues in evolution theory, this one refuses to go away. For despite the oft-repeated claim that no one has demonstrated repeatably the inheritance of acquired characters experimentally in the laboratory, the fact is that numerous researchers have done just that.

In the field of botany is the work of Alan Durrant of University College of Wales, Aberystwyth, who in 1962 induced changes in the flax plant by means of different kinds of fertiliser. Durrant bred some flax plants that were larger and heavier than the parent stock and another strain that was lighter and smaller. These trends persisted when the plants were bred in successive generations. The plant breeding was carried on for more than twenty years and it was shown that when the large and small plants were cross-bred, the offspring exhibited the Mendelian pattern of inheritance, proving that the change is genetic.[2]

The results have been replicated by J. Hill at the Welsh Plant Breeding Station, where a permanent change has been effected in the tobacco plant, *Nicotiana rustica*. In the case of the tobacco experiments, the flowering time was also changed.[3] Chris Cullis has reviewed all the work and has suggested a model to explain the induced changes in terms of molecular genetics.[4]

It may be said that these experiments are relevant to plants but not to animals. However, there is also experimental evidence from the animal world. As long ago as 1918, Guyer and Smith ground up the eye-lenses of rabbits and injected it into birds. When serum made from the birds' blood was injected into rabbits, their offspring were born with small or defective eyes, or with none at all, and the defects continued through a succeeding nine generations. The reason for choosing eye-lens tissue was that it is well known to provoke immune reactions.

The specific reason that Darwinian geneticists reject any form of Lamarckism is their belief that the genes are unalterably separate from the cells of the body and that there is no route by which changes could be communicated to them from outside. This belief was first enunciated by August Weisman in his 1893 book *The Germ Plasm: A Theory Of Heredity*.[5] It was re-stated as recently as 1970 by no less an authority than Francis Crick who, with James Watson, deduced the structure of the DNA molecule, together with the code by which it transmits genetic information. Crick said that genetic information could travel from DNA to protein but not from protein to DNA.[6]

Both Weisman and Crick have been shown to be mistaken by the work of Howard Temin at Wisconsin University in 1971. Temin discovered that viruses can transport genetic material into host cells and embed it in the host DNA where it will later replicate itself using the host cell's factory facilities for synthesising proteins. In order to perform this microbiological confidence trick, the viruses manufacture a special enzyme which Temin called reverse transcriptase and for the discovery of which he received a Nobel prize in 1976.[7]

Having found a two-way channel of communication between the genes and the outside world, science still lacked a mechanism by which the demands of the environment could directly affect the germ cells: how, say, the constantly stretching wading bird could transmit to its genes his desire to stay dry. Three years after Temin stepped onto the podium in Stockholm to shake hands with King Gustav, a young Australian biologist, Ted Steele, proposed just such a mechanism. In 1979, Steele proposed that mutations could occur in body cells, be copied to other body cells by viruses and finally be transmitted by viruses to the germ cells of the sperm in men or egg in women, and so become inheritable.[8]

The next problem was to design an experiment that would test Steele's idea. A colleague of Steele's, the Canadian Reg Gorczynski, neatly solved the problem by constructing an experiment merely by adding a new twist to the famous experiment of Sir Peter Medawar, who won a Nobel prize for showing that immune tolerance can be acquired from outside. Medawar's original experiment was concerned with a phenomenon that has become familiar to everyone in an age of organ transplants – that of rejection of tissue. The body's immune system will reject any cells that are not genetically identical and hence which it identifies as alien. Medawar showed that if alien cells are injected into a new-born mouse, then later in life it will accept a skin graft from the same source. His experimental success was graphically depicted in newspapers and magazines round the world with pictures of black mice having patches of white skin successfully grafted on and white mice equally at home with black patches.

Reg Gorczynski set out to duplicate Medawar's experiment

but to find out if the tolerance so established was inheritable. His experiment showed that 50 per cent of the offspring of the tolerant mice were also tolerant in the next generation and the grandchildren were tolerant in between 20 per cent and 40 per cent of individuals.[9]

On the face of it, this experiment successfully demonstrated the genetic inheritance of acquired immunity. It is fair to add that a team of distinguished scientists including Medawar himself and Professor Leslie Brent of St Mary's Hospital Medical School attempted to repeat Steele and Gorczynski's results and were unable to do so. The position at the moment is that the jury is still out. But regardless of the outcome of this particular experiment, it is no longer possible for neo-Darwinists to assert that outside agencies cannot communicate genetic changes via the mechanism of DNA replication. They can.

Additional, and very suggestive, evidence has come from two series of experiments conducted in the past three years in the United States. The first was conducted by British biologist Dr John Cairns and two colleagues at Harvard University in 1988. The second, a repeat of the Cairns experiments with tighter controls and extended objectives, was carried out by Dr Barry Hall of Rochester University in 1990. The experiments were conducted on bacteria, principally the species *Escherichia coli*. What they demonstrate is that when the bacteria are deprived of certain essential nutrients such as the amino acids tryptophan and cysteine, they are capable in this extremely hostile environment of giving rise to descendants capable of synthesising their own nutrients. What is taking place, believe Cairns and Hall, is that the bacteria are mutating and that the mutation is not random but internally directed by the needs of the organism in the direction of being able to synthesise the necessary nutrients.[10]

The only aspect of this experiment which it seems to me needs to be looked at carefully is the meaning which should be attached to the use of the word mutation in this context. The bacteria in the experiment are remaining the same species, *E. coli*. What is happening is that the individual bacteria comprising the test specimens that are defective at tryptophan production are dying because they are being deprived of external supplies of the nutrient, while those individuals capable of synthesising the material are flourishing. This is possibly reminiscent of the way in which the melanic form of the peppered moth prospered while the normal form died off, and might only represent a shift in population. It would, I think, take the appearance of a new species of bacteria altogether, or a function which did not previously exist, to confirm that mutation has taken place whether random or directed.

If these latest experiments are confirmed, it will almost certainly mean that we must look again much more closely at some form of Lamarckism. However unlikely it seems and however difficult it proves to obtain experimental confirmation, it looks increasingly probable that in some unknown way, individuals can not only adapt to their environment or way of life but can also sometimes pass on that adaptation to their offspring.

The aim of this chapter is to summarise the main alternatives to neo-Darwinism, of which some form of Lamarckism is possibly the principal contender. But it is by no means the only serious alternative proposal. Others include evolution by sudden jumps, the cause of which is uncertain, proposed by Richard Goldschmidt, Professor of Genetics at the University of California at Berkeley; the origin of life from space, as proposed by astronomer Sir Fred Hoyle; and a specialist variety of the extra-terrestrial hypothesis advanced by Francis Crick.

Some of neo-Darwinism's most important supporters have defected from the cause in recent decades and have espoused various heretical alternative ideas. The most prominent biologists to defect from the synthetic theory since the Second World War have been Richard Goldschmidt, who described the function of the chromosome, and C. H. Waddington, maverick Professor of Biology at Edinburgh University. The most recent heretics have included Harvard's professor of palaeontology Stephen Jay Gould, and his fellow palaeontologist Niles Eldridge, and British astronomer Sir Fred Hoyle. All have come in for vitriolic attacks from their former colleagues and fellow-believers and have had their ideas vilified and misrepresented. And all have dared to challenge the received wisdom of uniformitarian rates of change, and slow sure microscopic mutation coupled with blind chance.

Goldschmidt's concern has been to account rationally for the puzzling gaps in the fossil record by accepting them as real, rather than as inconvenient obstacles to an otherwise elegant theory. Goldschmidt coined the poignant and graphically descriptive phrase 'hopeful monster' to describe his heresy. His idea is simply that perhaps evolution proceeds by large jumps (known in the trade jargon of evolution as 'saltations'). Perhaps macroscopic mutations have occurred such that one day a reptile laid an egg from which hatched the first bird.[11]

This is actually no less probable than the idea that bears might mutate into whales. But the trouble with hopeful monsters is that they create a problem of exactly the same magnitude as the one Goldschmidt is trying to solve. To get from an ancestral reptile to a winged birdlike creature by conventional neo-Darwinist micromutations would take 100 or 1,000 or perhaps even 10,000 individual steps, each step representing a generation. Taking Sir Julian Huxley's rate of mutation (once in every million births) as a rough guide, then one would expect many millions – probably billions – of transitional individuals to have lived, at least some of which would be represented in the fossil record.

However, if viable macromutations occur (and there is no evidence for them, just as there is no evidence for beneficial micromutations) then most of them would be disadvantageous (perhaps fatal) to their carriers. Wings might conceivably be of assistance to a small lightweight reptile, but would be of little help to a 20–foot sea-going crocodile or a 200–foot brontosaurus. So there would have to be just as high a ratio of unsuccessful hopeful monsters to successful hopeful monsters as there would be transitional micromutations to stable species.

More simply, the fossil record would be littered with the bodies of one-off macromutations that did not work. For every macromutation like the hypothetical bird, there would be millions of one-legged crocodiles or aardvarks with wings. In fact, no one has recorded finding a single such failed monster. For Goldschmidt's idea to be correct, nature would need a 100 per cent success record. The difficulty this creates is that if nature possesses a mechanism that makes trial and error unnecessary, then the entire apparatus of random mutation and natural selection goes out of the window.

Stephen Jay Gould and Niles Eldridge of Harvard have proposed a theory of 'punctuated equilibrium', in order to account for the lack of fossil remains of transitional species. They have suggested that evolution is not a constantly occurring phenomenon; that species may have remained stable for long periods of geological history, leaving many fossil remains, and that the periods of evolutionary change, when they came, did not last for long. This would account for the lack of transitional fossils.[12]

The difficulty with punctuated equilibrium is that it is wholly speculative and has been introduced simply to account for the lack of fossils that ought exist in the neo-Darwinist theory. The ideas of Sir Fred Hoyle and his fellow astronomer Chandra Wickramasinghe on an extra-terrestrial origin of life are certainly guaranteed to liven up an otherwise dull winter's evening at your local pub. Surprisingly, their proposal is not new. In 1908 the distinguished Swedish chemist Svante Arrhenius suggested in his *Worlds in the Making* that living spores could be driven through space by the pressure of light from the stars.[13] Hoyle and Wickramasinghe's proposal is based on their belief that interstellar space is filled with clouds of dust consisting mainly of cellulose or similar sugar-like organic material. The comet Kohoutek was examined spectroscopically on its near approach to Earth in 1973 and was found to contain at least two organic molecules, methyl cyanide and hydrogen cyanide, along with rock dust, polysaccharides and related organic polymers, all possible building blocks for life. The two astronomers' idea also involves the idea of Earth colliding with a comet at some time in the past.[14]

The authors say, 'The Earth could have acquired all of its volatiles – including all the oceans – from such collisions [with comets]. And, of course, the presence of organic prebiotic chemicals such as we have discussed would have led to a vast input of life-forming materials to the Earth.'

Hoyle has made a powerful case for these ideas. In recent years several microorganisms have been discovered that can withstand the extremely hostile conditions of space. The bacterium *Micrococcus radiophilus* can survive exposure to X-rays at doses that would kill humans, while *Pseudomonas* has been found living quite happily in the core of an American nuclear reactor. Bacteria of the species *Streptococcus mitis* were inadvertently sent to the Moon in the unmanned Surveyor III in 1967 and were 'rescued' still alive two years later by the crew of Apollo 12 who brought back Surveyor's TV camera. The organism had been subjected to very low pressure and temperatures of minus 100 degrees Celsius.

Even more significantly, in 1981, Hans Dieter Pflug tentati-

vely identified microorganisms closely resembling the bacterium *Pedomicrobium* and a virus resembling influenza inside a
meteorite that fell in Australia in 1969. (As well as being amongst
the first to identify extra-terrestrial organisms, it was Pflug who
identified the oldest fossil on Earth, the 3,800-million-year-old
Isosphaera organism.)[15]

Francis Crick has made a further proposal. In his book *Life
Itself* he, too, suggests an extra-terrestrial origin for life but
believes that it is unlikely that organic molecules of any complexity could survive drifting in interstellar space. He suggests
instead that life in microscopic form may have been sent to
other planets by alien beings in suitable protective vessels; that
life is like a message in a bottle.[16]

If microscopic life originated elsewhere in the universe, then
it is still necessary to account for the evolution of life from the
microscopic to the macroscopic level. Extra-terrestrial origin
would assist neo-Darwinists to the extent that it relieves them of
accounting for the spontaneous synthesis of self-replicating
molecules on Earth, although, of course, it is still necessary to
account for their emergence on the planet of origin. The idea
also assists neo-Darwinists to the extent that it enables one to
conjecture a planet with conditions different from those on
Earth, and more favourable to the formation of life, perhaps
even with uniformitarian conditions of planetary development
and with the requisite billions of years of time available, which
we now know may not have been available on Earth.

At root, however, the same basic questions arise wherever life
is said to originate: what non-living mechanism can have given
rise to the first self-replicating cells and how; and what was the
mechanism of evolution from the cellular level to the present-
day plant and animal kingdoms?

Was it random mutation with natural selection? Punctuated
equilibrium? Or the hopeful monster? Or something else
entirely?

CHAPTER 18

The Facts of Life

Zoologist Bernard Heuvelmans once observed that just because a country is on the map, it doesn't mean that we know all about its inhabitants. Science has achieved miracles in the elucidation of the most complex microscopic structures and the furthest galaxies, yet there are deeper unsolved mysteries in the average suburban garden.

The swallowtail butterfly begins its life cycle by emerging from an egg as a caterpillar, enters a pupal or chrysalis stage and re-emerges as the familiar winged insect. While inside its pupa, however, the caterpillar undergoes a metamorphosis whose nature is not understood at all. The body of the caterpillar dissociates completely into an amorphous cellular liquid referred to as a 'soup'. The soup then reorganises itself into the structure of a butterfly.

To say that this process is not understood is not merely to say that BBC cameras have not so far been lucky enough to catch it on film. It means that no stage or aspect of this physical process can be accounted for or even guessed at with our current knowledge of chemistry, physics, genetics, or microbiology, extensive though they are. It is completely beyond us. We know practically nothing about the blueprint or programme governing the metamorphosis, or the organising agency that executes this plan.

In attempting to gather the strands of evidence from the natural world that might point the way to an alternative view of evolution, there seem to me to be three key kinds of observation, three persistently recurring themes that are crying for answers: the unerring accuracy of nature, her lack of trial and error; the presence of a systematic programme above the cellular level, controlling somatic development; and the overwhelming probability that environmental factors can in some unknown way directly affect the genetic structure of the individual.

The non-existence of transitional types (including failed monsters) in the fossil record and in the contemporary animal kingdom shows that nature goes unerringly to its target. The human eyelid exactly covers the human eye. The process that began the eyelid growing stopped it growing when it was the right size. It cannot be maladaptive to have an eyelid a little longer than needed – yet no creature has such an 'imperfection', in this anatomical detail or any of the myriad other details.

This is merely among the obvious examples of a universal phenomenon that we take for granted. The fact that a child's second teeth are adult-sized (even at age 7), the fact that little orange trees have little oranges and large orange trees large ones; the fact that the individual parts of any organism, from a tadpole to an elephant, are all in the correct relative scale.

Henry Williams of the University of North Carolina made an illuminating discovery at the turn of the century when he pressed sponges through a cloth until they were dissociated into individual cells. The cells spontaneously came together again and formed new sponges on their own initiative. In 1963, T. Humphreys confirmed and enlarged on Williams's discovery.[1] Johannes Holtfreter of the University of Rochester took up Williams's experiments after the Second World War and found that the cells from the embryos of vertebrates, too, will reassemble themselves when dissociated. In 1952 A. A. Moscona of the University of Chicago tried similar experiments with the tissues of chicks and mice. He found that dissociated kidney cells not only reassembled into kidney tubercles but also began to secrete kidney enzymes. Similarly, liver cells will reassemble into structures resembling the intact organ and carry out the liver function of accumulating glycogen. Heart cells – almost incredibly – coalesced into rhythmically contracting tissue.[2]

This sort of behaviour is also inexplicable at present, not I suspect because of some matter of detail, but because there is something big happening of which we know nothing as yet. It is not a matter of cells merely being attracted together like hydrogen atoms. These cells have a joint function which they cannot possess individually – like the heartbeat of heart cells. There is a programme being executed. How is it coded? Where are the instructions?

The many resoundingly pointless breeding experiments with the fruit fly *Drosophila* did throw up one highly illuminating discovery. To use the existing terminology, the fly possesses a mutant recessive gene (that is, one which normally plays no part in reproduction) which if present in both parents results in an offspring that is eyeless. If a stock of such eyeless flies is bred, then their offspring can only be eyeless too. Yet within a few generations offspring appear which do have normal eyes.

It would be absurd to imagine that nature has repeated in a few months what is supposed to have taken millions of years to occur: the origination of an eye by chance mutation. The ortho-dox explanation of this phenomenon is that the other genes have somehow 'deputised' for the missing gene by a recombina-tion. The significant point about the eyeless fly is that it again demonstrates some kind of programme control in action. The fly's genetic mechanism 'knows' that it lacks an important gene; and is able to take effective 'action' to compensate. The question is, where does this programme reside and how is it invoked and executed?

As long ago as 1895, German biologist Hans Dreisch per-formed an experiment that caused him to develop a whole new philosophy of biology. While working at the Zoological Research Station at Naples, Dreisch experimented on a sea-urchin egg. He killed half the egg but discovered that the remaining half developed into a perfectly normal embryo, except that it was half normal size. A rather similar experiment was conducted by B. I. Balinsky in 1951 when he transplanted tissue from the embryonic optic nerve of a large species of amphibian to the embryo of a smaller but related amphibian. The result was a perfectly formed eye with all parts in propor-tion, but intermediate in size between the two animals.[3] Once more, in both these cases, there is some kind of global supervi-sory function being exercised which seems to be 'aware' of an overall plan.

The example of the parallel evolution of the placental mam-mals of Europe and the marsupial mammals of Australia, des-cribed in Chapter 14, is further evidence for some overriding principle at work. Neo-Darwinists believe that a single shrew-like ancestor has independently evolved into carbon copies of wolves, cats, rats, and a dozen other mammals on widely separated continents, simply by virtue of their common lifestyles.

But it is plain that if the shrew-like creature of the Cretaceous really is the ancestor of both marsupials and placentals, then its evolutionary 'trajectory' has been strictly circumscribed by natural laws, just as the flight of a cannonball is circumscribed by gravity. The repertoire of options open to evolution has been dictated by a strategic plan or programme. Where does that programme reside? How is it executed? What is the 'gravity' of evolution?

Earlier on I referred to computers and their programs as a fruitful source of comparison with microbiological genetic processes since both are concerned with the storage and reliable transmission of large quantities of information. Arguing from analogy is a dangerous practice, but there is one phenomenon connected with computer systems that I believe may turn out to be of some importance in understanding biological information processing strategies and that I feel is worth elaborating on.

The phenomenon has to do with the computer's ability to refer to a master list or template and to highlight any exceptions to this master list that it encounters during processing. This 'exception reporting' is profoundly important in information processing. For instance, this book was prepared using a word-processing program that has a spelling checker. When invoked, the spell checker reads the typescript of the book and compares each word with its built-in dictionary, highlighting as potential mistakes those it does not recognise. Of course, it will encounter words that are spelled correctly but are not found in a normal dictionary – such as 'deoxyribonucleic acid' – but the program is clever enough to allow me to add the novel word to the dictionary, so that the next time it is encountered it will be accepted as correct instead of reported as an exception – as long as I spell it correctly.

In other words, the spelling checker isn't really a spelling checker. It has no conception of correct spelling. It is merely a mechanism for reporting exceptions. Using these methods, programmers can get computers to behave in an apparently intelligent or purposeful way when they are really only obeying simple mechanical rules and, not unnaturally, this gives Darwinists much encouragement to believe that life processes may at root be just as simple and mechanical.

In cell biology, there are natural chemical properties of complex molecules that lend themselves to automatic checking and excepting of this kind. For example many molecules are stereospecific – they will attach only to certain other specific molecules and only in special positions. There are also very much more complex forms of exception reporting, for instance as part of the brain's (of if you prefer, the mind's) cognitive processes, as when we see and recognise a single face in the crowd or hear our name mentioned at a noisy cocktail party.

In the case of the spelling checker, the behaviour of the system can be made to look more and more intelligent through a process of learning if, every time it highlights a new word, I add that word to its internal dictionary. If I continue for a long enough time, then eventually, in principle, the system will have recorded every word in the English language and will highlight only words that are indeed misspelt. It will have achieved the near-miraculous levels of efficiency and repeatability that we are used to seeing in microbiological processes. But something strange has also been happening at the same time – or, rather, two strange things.

The first is that as its vocabulary grows, the spelling checker becomes *less* efficient at drawing to my attention possible mistakes. This unexpected result comes about in the following way. Remember, the computer knows nothing of spelling, it merely reports exceptions to me. To begin with, it has only, say, 50,000 standard words in its dictionary. This size of dictionary really only covers the common everyday words together with a modest number of proper nouns (for capital cities, common surnames and the like) and doesn't leave much room for unusual words. It would, for instance, include a word like 'great' but not the less-frequently used word 'grate'.

The result is that if I accidentally type 'grate' when I really mean 'great', the spell checker will draw it to my attention. If, however, I enlarge the dictionary and add the word 'grate', the spell checker will ignore it in future, even though the chances are that it will occur only as a typing mistake – except in the rare case where I am writing about coal fires or cookery.

One can generalise this case by saying that when the dictionary has an optimum size of vocabulary, I get the best of both worlds: it points out misspellings of the most common words and reports anything unusual which in most cases probably will be an error. (Obviously to work at optimum efficiency the size of dictionary should be matched to the vocabulary of the writer.) As the dictionary grows in volume it becomes more efficient in one way, highlighting only real spelling errors, but less efficient in another: it becomes more probable that my typing errors will spell a real word – one that will not be reported – but not the word I mean to use. Paradoxically, although the spelling checker is more efficient, the resulting book is full of contextual errors: 'pubic' instead of 'public', 'grate' instead of 'great', and so on.

It requires a human intelligence – a real spelling checker, not a mechanical exception reporter – to make sure that the intended result is produced.

I said two strange things have been happening while I have been adding words to the spelling checker. The second is that on the odd occasion, the system has highlighted a real spelling mistake to me – say, 'problm' instead of 'problem' – and I have mistakenly told the computer to add the word to its dictionary. This, of course, has the very unfortunate result that in future it will cease to highlight a real spelling mistake and will pass it as correct.

Under what circumstances am I most likely to issue such a wrong instruction? It is most likely to happen in the case of words that I type most frequently and that I habitually mistype. Anyone who uses a keyboard every day knows that there are many such 'favourite' misspelt words that get typed over and over. Once again, only a real spelling checker, a human brain, can spot the error and correct it.

The reason that the computer's spell checker breaks down under these circumstances is that the simple mechanisms put in place do not work from first principles, do not work in what electronics engineers call 'real time' (they are not in touch with the real world), and do not employ any real intelligent understanding of the tasks they are being called on to perform. So although the computer continues to work perfectly as it was designed to, it becomes more and more corrupted from the standpoint of its overall function.

I believe that this analogy may well have some relevance to Darwinists' belief that biological processes can at root be as simple as the spelling checker. It is easy to think of any number of simple cell replication mechanisms that rely on exception reporting of this kind. I believe that if biological processes were so simple, they too would become functionally corrupt unless there were some underlying or overall design process to which the simple mechanisms answered in a global way, and which were capable of taking action to correct mistakes. This is the mechanism that we see in action in the case of the 'eyeless fly', *Drosophila*; in Dreisch's experiment with the sea urchin and Balinsky's with the eyes of amphibians; and the 'field' that governs the metamorphosis of the butterfly or the re-constitution of the cells of sponges and vertebrates.

Darwinists believe that the only overall control process is natural selection but the natural selection mechanism could not account for the cases referred to above. Natural selection works on populations, not individuals. It is capable only of making sure that creatures with massively fatal genetic defects die in infancy, or that populations that are geographically dispersed will eventually produce sterile hybrid offspring. It is such a poor feedback mechanism in the sense of exercising an overall regulating effect that it has failed even to eliminate major congenital conditions such as Down's syndrome. Because natural selection offers only death or glory it cannot provide the microscopic adjustments that the individual needs. Yet we are asked to believe that a mechanism of such crudity can creatively supervise a program of gene mutation.

This is plainly wishful thinking. The key question remains: what is the location of the supervisory agency that oversees somatic development? How does it work? What is its connection with the cell structure of the body?

Whether they are Darwinists or vitalists, biologists have begun to talk in terms of 'morphogenetic fields'. D. J. Pritchard, a Darwinist geneticist from Newcastle University, wrote in 1990:

There is a great deal of evidence that organs and organisms have an awareness of their 'wholeness' (Dalq, 1951; Lillie, 1927; Spemann, 1924) such that when a portion of the whole is lost, steps are taken to replace it. For example salamanders will regenerate their limbs (French, Bryant and Bryant 1976; Wallace, 1981); if a sponge is disaggregated into single cells these will reaggregate to form a perfect sponge (Humphreys, 1963). Embryologists recognise 'morphognetic fields' which have spatial unity with respect to the organisation of their constituent parts. If a field is divided into two a complete structure can form in each half independently of the other. Our own retinas began as the two halves of an initially single retinal field. If division of the retinal field fails the result is a single, central eye, a condition known as cyclopea. What evolution has created within the bodies of animals are integrated, self-organising systems which are not just defined by their component parts, but actually define those components.[4]

These tantalising glimpses of the unity of organic structures is as far as our present experimental knowledge takes us. Only further experiment and a certain amount of luck can provide the hard data that will solve fully these baffling questions and we must await the acquisition of new facts. In the absence of concrete answers, I would like to offer some speculations.

To begin with, we have a working hypothesis in Ted Steele's proposal that viruses are able to replicate mutations in somatic cells and transfer them to sexual cells, where they become inheritable. The next question to be asked is, what kind of cellular changes might be induced in somatic cells? And, exactly how might they be induced?

C. H. Waddington – an unusual combination of an academic with an anarchic sense of humour – has essayed just such a mechanism. It must be said that Waddington dreamt up this mechanism in a light-hearted vein simply to infuriate orthodox neo-Darwinists (especially Jacques Monod of the Pasteur Institute who had accused Waddington of being a Lysenkoist – an even worse crime than Lamarckism). In his essay *How much is evolution affected by chance and necessity*, Waddington includes a massive footnote outlining his idea.[5] In a crude and simplified form it is this. It has been established that important parts of the DNA molecule are repeated many times in the chromosomes – rather like back-up tapes. Just like back-up tapes, these replicate versions may vary slightly. There is also another set of tapes in the form of mitochondrial genes, which are further structures in the cell. All these genes are closely involved with the important metabolic processes that go on within the cell. It is thus not inconceivable that the rates of multiplication of slightly differing genes would be influenced by the particular metabolic circumstances reigning in the cell in question. And it is not inconceivable that the imposition of certain metabolic conditions on an organism might change the proportion of variant forms of gene within the population (of all the back-up copies) to be passed on to the next generation. The effect of this would quite simply be the direct inheritance of an acquired character.

Put more simply, the metabolic stresses placed by an individual on his cellular structure might determine which tape is selected from the library for duplication. To try to add a little colour to the hypothetical mechanism (which Waddington himself did not) one can imagine a very athletic woman stimulating the metabolism of her cells such that replicate DNA sequences coding for physical agility are promoted preferentially with the result that she has athletic daughters.

Waddington called his idea an 'outrageous speculation'. What he may not have known when he dreamt it up in 1974 is that he had only to account for differential multiplication of the DNA replicates in ordinary body cells: for Steele's viruses could replicate the chosen DNAs to the sexual cells through 'reverse transcription'. This could make his suggestion hundreds if not thousands of times less outrageous and more probable.

The hypothetical example given above would be an example of a physical behaviour affecting somatic cells. Are other forms of influence possible? The answer appears to be that psychological states may also affect somatic cells. Epidemiologists believe they have identified a 'cancer personality' – a set of individual character traits which, if possessed mainly or wholly by one individual, may predispose that person to cancerous illness: that is, the faulty replication of somatic cells. If it is true that personality factors can affect cell biology, and if viruses can copy genetic mutation from somatic cells to sexual cells, then it follows that personality factors could in principle be inheritable. To carry this speculation a step further, some of the personality traits which compose the 'cancer personality' are psychological rather than physical (for example excessive anxiety). This raises the possibility that purely psychological factors could be translated into both somatic and ultimately genetic factors: the content of an individual consciousness could affect his or her body and the bodies of any offspring.

Presumably, the metabolism of the 'cancer personality' is actually different from that of a non-cancer personality in some distinctive way. For instance, anxiety alters the balance of some hormone or enzyme which ultimately results in alteration to somatic cells. If so, the nature of those differences may hold an important key for biology.

One further possibility – perhaps a rather worrying one – remains to be explored. For more than fifty years, it has been recognised that, at the nuclear level, our solid world dissolves into a cloud of fuzzy probabilities. Until recently, lip service was paid to the principle of uncertainty in physics, but no serious scientist would care to admit that he had designed an experiment taking himself into account. Now a concrete experimental result has been obtained which clearly shows the influence of the observer at the quantum level.

The National Institute for Standards and Technology in Colorado reported an atomic experiment in *Physical Review* in March 1990 in which the result was determined by the observer.[6] The NIST experiment involved heating with radio waves a container of beryllium atoms and measuring the ratio of isotopes formed. The radio pulses convert the atoms from one isotope to another. The researchers used a laser beam to show up the results since it would cause atoms in their original state to emit light but not atoms in the altered state.

What they found was that the more measurements they made with their laser beam, the greater the number of atoms that remained unaltered. The very act of observing stopped them changing state, regardless of the effect of the radio pulses. Notice that this is not simply a matter of the laser beam preventing the experiment from progressing or directly interfering with the changes in atomic state. The explanation is that observing a particle causes it to collapse from a fuzzy probabilistic cloud into a definite mass at a definite point in space and time, as predicted by quantum mechanics.

The question this experiment raises is, if merely observing an event causes changes to occur at the atomic level, and if genetic coding is controlled by atomic structures, can genetic mutation be caused by direct influence at the quantum mechanical level?

Is it even conceiveable that as, Hans Dreisch conjectured, 'The mind may carry out a morphogenetic action at a distance'? Can we wish for wings and get them? Probably not. Does a healthy mind promote a healthy body? Almost certainly. Is there anything in between? Who knows?

Thomas Huxley, Darwin's champion, observed that the great tragedy of science is the slaying of a beautiful idea by an ugly fact. Darwin's original conception was a beautiful idea. It seemed to offer an elegantly economical solution to the greatest mystery of all: the origin of life on Earth and the descent of mankind. Sadly, it has received too many mortal blows from the ugly facts of scientific enquiry to remain viable.

The prospect of facing the future without neo-Darwinism is not an attractive one. Its demise will leave a yawning gap in the life sciences and historical geology with no obvious successor theory; a hole that it is impossible to imagine being filled by any current competitor. How has life evolved if not by chance?

It is the customary fate of one who delivers the fatal stroke to be called upon to replace the deceased theory with a better one. This is thoroughly illogical, quite unfair and perfectly understandable. While I do not possess an alternative theory in my back pocket, I may at least suggest what kind of new theory it might be, and where it might be found.

The alternative mechanisms so far discussed in this chapter have in common that they all approach the problem from the accepted premises of classical science – indeed in a way in which Darwin himself might approach it if he were alive today and possessed of today's body of scientific knowledge. I have a deep-rooted suspicion, however, that the real solution may be found in adopting a quite different approach; that the natural phenomena that may well provide an explanation of the origin of species are at present so imperfectly understood that they have baffled those physicists who have bothered to examine them at all, and have been ignored almost entirely by biologists.

For most of this century, physics has had to accept the indignity of a principle of uncertainty. Physicists have been compelled to drop their neat logical picture of the universe as a great machine, and their unambiguous, clockwork model of the atom. In the place of these certainties, physical scientists have been obliged to put intangible, unimaginable abstractions. Instead of billiard ball particles like electrons, there are probability waves. Instead of matter composed of particles and energy composed of waves, there is light made of particles, and objects made of matter waves. In this surreal sub-atomic world, matter has ceased to have any solid form and has no more than a tendency to exist.

While these turbulent events have been taking place in the physics faculty, down the corridor in the biology department it has been business as usual. Biologists have made remarkable discoveries, but they all have the familiar nineteenth-century hallmarks of clockwork certainty. Deducing the structure of the DNA molecule is a brilliant scientific achievement. But the blue and red ping-pong balls of the molecular model remain frustratingly incapable of telling us what life is.

Using the mechanistic, reductionist approach of Victorian science, biology has not so much explained life as explained it away. The body is a machine, a matter of chemistry and electricity. Thought is merely a by-product of the computer-like brain which pulls the body's levers. Evolution is no more than a marriage of chance and chemistry. There is no ghost in the machine: man *is* the machine. It is out of this Frankenstein approach that neo-Darwinism was born and is sustained: by the science of Mendel and Kelvin, rather than that of Heisenberg and Planck.

As we near the end of the twentieth century, I believe that biology, too, will be compelled to drop its mechanistic approach and recognise that chemistry and statistics alone will not explain the nature of life. The absurd and baffling world of the nuclear particle is beckoning those in the life sciences as it beckoned physicists decades ago. Biologists are, as it were, hesitating on the shores of an unexplored continent. What they will find when they venture inland it is impossible to say. But it is possible to gain some clues from the discoveries that have been made by their colleagues from the physics laboratory who set off some fifty years ago and have a substantial head start.

If the new physics has a central idea to sustain it, it is that of wholeness. In 1935 Albert Einstein, Boris Podolsky and Nathan Rosen set their colleagues in physics a baffling conundrum. Trying to answer the question of whether quantum mechanics really tells us anything about the nature of the physical world, the three physicists proposed a thought experiment with astounding consequences. They showed theoretically that atomic events which appear to us as separate must in fact be connected in some unknown way. And moreover, that such events can communicate information to each other instantly – faster even than the speed of light which is thought to be a limiting velocity in the physical world.[7] The three physicists predicted that whatever happened to a nuclear particle would be reflected in the behaviour of its twin particle in a closed system, regardless of where they were. Even if they were billions of miles apart, a change in the momentum of one particle would be mirrored in its twin instantly – as though the particles were able to communicate their experience instantaneously.

Einstein, who doubted that quantum physics gave a real description of real events, thought it more likely that the twin particles were behaving in a way that merely appeared to be co-ordinated in a cause and effect manner because they were both obeying some third, hidden factor affecting them both, a factor known to physics as a local hidden variable. He, and most physicists, preferred this explanation because they do not like to have to draw upon any form of inexplicable action-at-a-distance. In any case, it was thought that even if the extra-ordinary connectedness of nuclear particles was real, it was an effect which existed only at the nuclear level – not in the real world of tables and chairs and certainly not in the realms of biology.

In recent decades a number of research groups have conducted physical experiments which have confirmed the unlikely prediction of the Einstein-Podolsky-Rosen paradox. In 1972 Stuart Freedman and John Clauser at Berkeley performed an experiment which confirmed that photons – the quanta of light – really are mysteriously correlated. It is no mere philosophical contrivance to get physicists out of a conceptual difficulty; this wholeness or hidden connectedness is real. Even more significant, its effects can be felt at the macroscopic level, at the level of the everyday world including that of biology.

David Bohm, Professor of Physics at Birkbeck College at the University of London, has written of this connectedness in his book *Wholeness and the Implicate Order*.[8] Bohm sees the cosmos as a connected whole which he terms the implicate, or enfolded universe. The fragments of it that we perceive with our human minds and senses he terms the unfolded or explicate world. We see and understand only a tiny fraction of the underlying connected whole – the tip of the cosmic iceberg, as it were.

So far, few biologists have ventured to abandon the conventional viewpoint of the nineteenth century in favour of this strange new world their colleagues have discovered. But one researcher who was far in advance of his fellow biologists was Hans Dreisch, who, as mentioned earlier, concieved a vitalist theory of biology following his experiments with sea urchins. Dreisch concluded that the development of organisms is directed by 'a unifying non-material mind-like something . . . an ordering principle which does not add either energy or matter' to the processes it directs.[9] He suggested also that this principle might exist outside the normal framework of time and space – an idea strongly reminiscent of David Bohm's implicate order and the 'extra-curricular' connectedness of the Einstein-Podolsky-Rosen experiments.

Few of his fellow biologists shared Dreisch's view of nature. One exception is Sir Alister Hardy, Professor of Zoology at Oxford from 1946 to 1963. In 1949 Hardy astonished the British Association for the Advancement of Science by suggesting in his presidential address to the zoological section that telepathy was relevant to biology. In the *Journal of the Society for Psychical Research*, Hardy wrote: 'Assuming the reality of telepathy . . . the discovery that individual organisms are somehow in psychical connection across space is, of course, one of the most revolutionary . . . ever made.'

Hardy professed himself to be a Darwinist, but it was a strange variety of Darwinism which enabled him to assert that 'there is a general subconscious sharing of a form and behaviour design, a sort of psychic blueprint between members of a species,' and that 'the mathematical plans of growth seem to have all the appearance of a pattern outside the physical world which has served as a plan for selective action by way of changing combinations of genes.'[10]

The dangerously heretical ideas and experiments of zoologists like Dreisch and Hardy were not so much ignored by their fellow biologists as mentally quarantined, in case they should prove contagious. In the event, they have proved to be every bit as infectious as feared.

In the baffling new world of modern physics, scientists find themselves observing and examining a cosmos which has become less and less like a clockwork machine and more and more like an intelligence. Whether the intelligence is that of ourselves, the observers, or that of the world we examine is not yet clear and perhaps may never become clear. But it would surely be absurd to bestow intelligent characteristics upon the behaviour of nuclear particles but fail to accord such characteristics to living structures.

An Evolutionists' Apocrypha

CHAPTER 19

On Being Thick-Skinned

When critics of synthetic evolution assemble their evidence and put their case, it is usual for them to torture Darwinists by throwing at them inexplicable and complex individual examples of structures and behaviour from the animal and plant world which defy probability. The ammunition at their disposal is immense and creationists in particular never miss an opportunity to hurl an example or two at evolutionists rather as zoo visitors throw buns to the elephants. The practice is strongly resented by Darwinists and arouses deep feelings – Richard Dawkins confesses to 'despising' scientific creationists and what he calls their 'journalistic fellow travellers'.

I have manfully resisted the temptation to indulge in this amusing sport while developing my main arguments, and have avoided the seductive lure of resting any part of my case on Paley's argument from design, on the improbability of this or that anatomical feature, such as the complexity of the human eye, since I feel that these examples are as likely to cloud the issues as to clarify them. And most of the examples are beginning to become dog-eared from being hurled at Darwinists so often.

But since this book is an attempt to present a global critique of neo-Darwinism, it would be negligent of me to omit an entire body of evidence. So in the interests of completeness (and a little modest entertainment) I present the following Golden Treasury – or perhaps *grimoire* – of evolutionary impossibilities.

Darwinists have understandably had to become thick-skinned about such examples being thrown at them. But that is not the origin of the present chapter heading. Instead it refers to the first and, some think, most bafflingly improbable mutation – the thick skin on the soles of our feet. This thicker skin does not appear after birth as a result of walking around, but is present in the human embryo (and the embryos of some other species such as apes). It is thus an inherited characteristic. How does it come about that we have thick skin just where we need it and nowhere else on our bodies? The neo-Darwinist solution is that it is the result of a chance mutation. Presumably other human ancestors had chance mutations that gave them thick skin elsewhere – on their noses perhaps or their ears – but this did not increase survival chances and hence was not selected for.

Other species also have thickening of the skin in places uniquely suited to their mode of life: the African warthog has callosities on its wrists and forelegs on which it leans while feeding; the camel has them on its knees; that curious bird the ostrich has them back and front on its under side where it squats. All are inherited characteristics. All are present just where the animal needs them and nowhere else. No species is known which possesses unnecessary callosities. Does anyone *really* believe this is the result of random chance?

The human eye is generally taken as the archetypal 'impossible' structure. It is the one most often discussed and the one Darwin himself confessed gave him 'a cold shudder'. In one sense, the eye ought not to give evolutionists the shivers because it is only another structure – admittedly many times more complex than an arm or a wing, but degree itself is no objection to the principle of random mutation. Once you have accepted that mutation coupled with natural selection can produce something as complex as a DNA molecule or a bacterium, then it is no more than a matter of time before something as complex as the eye arises. And evolutionists have allotted themselves practically unlimited time. But it is not the complexity of the eye itself that causes Darwinists their difficulty. It is the problem of demonstrating all the many stages of the eye in transition. Consider this statement from Garrett Hardin:

> Were all other organisms blind, the animal which managed to evolve even a very poor eye would thereby have some advantage over others. Oysters have such poor eyes – many tiny sensitive spots that can do no more than detect changes in the intensity of light. An oyster may not be able to enjoy television, but it can detect a passing shadow, react to it as if it were caused by an approaching predator, and – because it is sometimes right – live

another day. By selecting examples from various places in the animal kingdom, we can assemble a nicely graded series of eyes, passing by not too big steps from the primitive eyes of oysters to the excellent (though not perfect) eyes of men and birds. Such a series made up from contemporary species, is not supposed to be the actual historical series; but it shows us how evolution could have occurred.[1]

This view is echoed by Sir Gavin de Beer, an embryologist and Director of the British Natural History Museum, in his *Atlas of Evolution* where he illustrates a sequence beginning with the primitive eye-spot and culminating in the eye of mammals. 'There can be little doubt,' he wrote, 'that the series of stages described through which the eye passes in embryonic development is a repetition of the manner in which it evolved.'[2]

This is a fair summary of the neo-Darwinist view. But the difficulty with Hardin's argument is that it specifically fails to do what he sets out to do – to demonstrate step by step the evolution of the human eye. It says, in effect, that all the species in the living world today have evolved by random mutation and natural selection: they exhibit various kinds of eye from primitive to advanced; therefore the human eye has developed by such evolutionary stages. Hardin has reached his conclusion only by including it in his premises.

The fact that an oyster has a primitive eye does not demonstrate that complex eyes evolve from primitive eyes – that is the very matter in question. If palaeontologists could produce a series of fossil mammals, or reptiles or fish showing the eye in these various stages, their case would be made. But of course, if they could produce such a series of fossils, they would not need to concern themselves with medieval debates about eye complexity – they would already have made their case.

Interestingly, Professor Wolsky in his book *The Mechanism of Evolution* points out that light-sensitive organs in all creatures seem to have evolved in the places where light falls most intensely, suggesting that this appears to be some form of design.[3] The Darwinist's traditional response is that mutations that cause eyes in the 'wrong' place would not be adaptive and hence would not be selected for. But here they are attempting to have their cake and eat it, for they also argue that (in Hardin's words) 'even a very poor eye would have some advantage'.

Hence we should expect to find creatures with eyes in less than optimum locations, such as on the flanks or the base of the spine. But no such creatures exist, either today or as fossils.

My next example is not so much concerned with the evolution of new organs as the disappearance of existing ones. Evolutionists believe that, for example, the snake is a reptile which was originally like a lizard, but has lost its arms and legs as a result of adapting to a crawling mode of life. Similarly, the whale is believed by evolutionists to be a mammal which has returned to the sea, and lost its limbs in order to become streamlined for swimming. Despite the whale's enormous size, its thighbone has now shrunk to a mere eighteen-inches long and is on its way to vanishing entirely.

The question is, what was the evolutionary advantage of the thighbone becoming any smaller than the whale's streamlined body envelope? What was the evolutionary advantage of the snake's arms and legs disappearing altogether? Or the mole's eye sockets being filled with muscle? Is it really rational to suppose that *random* mutations appeared which progressively diminished just these organs until they vanished entirely, long after any survival advantage could have been gained? The concepts of mutation and selection are both flawed in explaining the whole field of regressive organs. It seems clear that some systematic process or programme is taking place which, once initiated, proceeds to a conclusion. Where does the 'programme' reside? How does the 'system' know when to start and stop?

One category of impossible mutations has to do with precision engineering: engineering to limits that we would find extremely difficult to emulate. The oft-quoted eye is in fact not very precisely engineered: its elements can vary by a substantial margin and the eye will still function reasonably well. Some natural structures, though, require an accuracy of millionths of a centimetre. The silvery skin of fish is designed to provide a reflective surface that enables them to remain camouflaged and unnoticed by predators, in the greenish gloom of the sea. To achieve this, the fish secrete millions of tiny nitrogenous crystals in layers on their skin and scales. But this is not all. To increase the efficiency of their reflective coating (from about 25 per cent reflective to as much as 75 per cent) the fish secrete multiple layers of mirror crystals sandwiched between layers of cell tissue. But to be effective, the layers have to be arranged exactly one-quarter of the wavelength of the incident light apart. For the greenish light of the undersea world, this means a separation of seven millionths of a centimetre.[4] Does anyone really believe that this precision was achieved by random mutation?

My personal favourite among the specimens in the black museum of incredible mutations is the general matter of the alternation of generations. This is seen for instance in jellyfish which reproduce by releasing eggs and sperm into the sea. The fertilised egg does not develop into another jellyfish straight away but settles down to another form of life as a flower-like polyp anchored to the sea bottom or a rock. Ultimately the polyp buds (in a different way from its parent jellyfish) and the buds grow into free-swimming jellyfish once more. In some types, the majority of time is spent in the free-swimming form with only relatively short spells as a polyp. The common sea anemone, on the other hand, spends most time anchored to rocks and little as a free swimmer.

The alternation of generations raises all kinds of fascinating questions concerning the adaptive advantage of such a way of life, and how it could have come about by microscopic mutations (it is hard to imagine how the alternation of generations could come about a little at a time – indeed this is one of the examples that made Richard Goldschmidt conceive his hopeful monster theory).

The aspect that fascinates me most is the fact that there is some kind of counting or timing mechanism at work here: a mechanism that recurs in animal and plant life. A few random examples will explain. The artichoke plant, grown by gardeners for its fruit, will crop for three years; the plant then dies, or sometimes lives on but will crop no more. However, if a cutting is taken and planted, it too will crop for three years. The common variety of asparagus crown will crop for seventeen years and then cease. Human children have two sets of teeth: the first set come through in a miniature size appropriate to a child; the second teeth come through full grown at adult size even though they usually appear at around only seven years. There is a species of bamboo tree that flowers every 117 years, and cacti that flower every twelve years. The ptarmigan and the arctic fox assume a whitish coat in winter and a brownish one in summer.

How does the sea anemone 'know' it is time to become a jellyfish? How does the artichoke 'know' its three years are up? How do the child's teeth 'know' they are second teeth and must be bigger than the first? How do they know what scale to be on at all? How do the ptarmigan and arctic fox 'know' when to change coat? And when to change back again?

The answer may be a relatively simple matter of genetic coding. For example, adult-sized teeth may be the product of a genetically coded scale factor that is applied to every protein synthesis regardless of its function in the body. But it is very hard to see how a timing mechanism can operate – especially across the generations – without some global or systematic function being invoked that controls the entire organism in some way. And this is specifically what Darwinists say does not exist.

There is one further important question relating to the issue of timing and that is the question of the storage of mutations. Suppose that the environment changes in a radical way and only some new adaptation can survive in the changed environment: for instance picture a lake which is drying out and the advantage that a fish with lungs will have over a fish that breathes through gills. When, exactly, does the new feature – the lungs – appear? Clearly, it would be cutting things a little fine to wait until the last pool is drying out and then hope that random mutation will provide a means of breathing air. It is necessary to assume that the lungfish is pre-adapted to the change. In this case, it is possible to argue that the lung provides a useful flotation device even before the lake dries up. But the important point to notice is that this pre-adaptation has to be true of every single one of the millions of evolutionary changes that are said to have taken place throughout the entire animal and plant kingdoms: they must all be in place before they are needed; not just lungs but feathered, articulated wings and protective camouflage and all the rest. If the fish's reflective coating is off by even a millionth of a centimetre, it will not reflect the right wavelength of light and could make the fish stand out, rather than remain hidden.

The improbability of the precise adaptation either occurring at the right moment, or being an existing adaptation that can luckily be further extended, has led neo-Darwinists to look for other explanations. They have latched on to the unused portion of the DNA in the chromosomes as being the answer. As mentioned earlier, only some 10 per cent of the DNA in the chromosome actually does anything. The other 90 per cent seems very similar, but does not actually cause any proteins to be synthesised and has been referred to as 'nonsense DNA'. Neo-Darwinists believe that this unused 90 per cent contains latent and recessive genes – mutations which exist in the gene pool of the species but have not so far been made operational because they offer no advantage. Each time an individual is born exhibiting one of these mutations, he or she dies, probably in infancy. Neo-Darwinists believe that if one day an individual were born with gills and the world were flooded, then he and his offspring would flourish while the rest of us die off. But the implication of this proposal is that we should see, in all species throughout the animal and plant kingdoms, the continuous emergence of such mutations – for if they do not emerge they can never be implemented. And this is precisely what we do not observe.

Notice that this is not merely another way of saying that we do not see transitional forms in the fossil record. These mutations would be happening all the time but *not going any-where*. They would be unique individuals, not transitions. For that reason they would be rare, but the point is that for nature to have available the raw material for a macromutation when an ecological crisis arises, there would have to be millions or even billions of such freaks. Presumably many of them would be the same sort of 'mistake' cropping up time and time again. But they do not exist either in life or in fossil form. Because mutations are not observed, we are entitled to conclude that the unused 90 per cent of the DNA molecule is not the home of beneficial mutations.

In 1940 Richard Goldschmidt felt concerned enough about the conventional neo-Darwinist view to throw down this challenge:

I may challenge the adherents of the strictly Darwinian view . . . to try to explain the evolution of the following features by accumulation and selection of small mutants: hair in mammals, feathers in birds, segmentation of arthropods and vertebrates, the transformation of the gill arches in phylogeny including the aortic arches, muscles, nerves, etc.; further, teeth, shells of molluscs, ectoskeletons, compound eyes, blood circulation, alternation of generations, statocysts, ambulacral system of echinoderms, pedicellaria of the same, cnidocysts, poison apparatus of snakes, whalebone, and finally primary chemical differences like haemoglobin versus haemocyanin, etc.

Goldschmidt adds that corresponding examples from the plant world could also be given.[5]

So far as I am aware, no Darwinist has accepted Goldschmidt's challenge. But whereas he was regarded as having a screw loose in 1940, he is taken a great deal more seriously today.

There is one further matter that I feel belongs among these apocrypha, rather than the seriously argued part of this book. It is not an 'impossible' mutation, although it is an old chestnut. It concerns the most glaringly obvious gap in the entire fossil record: the gap right at the beginning in the rocks of the pre-Cambrian period. In those rocks – some 5,000 feet of fine-grained sandstones apparently lying in unbroken succession – there are fossils of the earliest life forms: mushroom-like structures called stromatolites which were colonies of green algae. In rocks of the same age are the fossils of corals, sponges, snails, crabs, sea urchins, jelly-fish, trilobites and graptolites. There is no gradual appearance of more and more complex forms: the rocks contain a mature population of a vast range of creatures, many of which are indistinguishable from species which are alive today, together with a huge number which have since become extinct.

I feel it is a little unfair to throw this particular chestnut at evolutionists, because it is perfectly possible, as they claim, that there may be a gap in the geological sequence at this point and hence the evidence is missing (the same claim they advance for all the other 'gaps' in the fossil record). It is also possible that the precursors of these animals with hard shells and carapaces had only soft bodies and left no traces. It is, though, an extraordinary coincidence that this key section of the geological record should be 'missing', and coincidence is a card they have played once too often for it to remain convincing in explaining gaps.

In this book I have presented arguments that rest on empirical evidence as far as possible and have deliberately tried to avoid using arguments which depend solely on logic for their force. I am well aware that only nature is the arbiter of what is true and what is false, and that human speculation is too often defective however impeccable its logic. A well-known case of this sort is the inability of the bumble bee to fly. The bumble bee cannot fly because its wing area is insufficient in relation to its body weight; because it lacks an aerofoil section wing; and because its flat wing is not articulated in two planes – it merely flaps up and down on a simple hinge, generating an upthrust and a downthrust that are equal and opposite. The bumble bee, of course, remains unaware that it is defying all the laws of aerodynamics and continues to flit from flower to flower blissfully unconscious of its theoretical shortcomings.

However, there is one area of the synthetic theory that I feel is open to question because it is itself constructed on a purely logical basis and appeals to us for acceptance on the grounds that it seems to be self-evidently probable. This is the central issue of cumulative development by natural selection.

Darwinists say that the reason one species of animal disappears in the fossil record and is replaced by an apparently related but distinctively different species is because a beneficial mutation has occurred fitting the original creature better to its environment and the new mutated species eventually colonises the ecological niche more efficiently at the expense of the old. The old species then dies out because it has a new and more successful competitor in the form of its own mutant descendant. This process means that future mutations will take place in the gene pool of the new species, and thus cumulative evolution proceeds.

To take a specific example, it is assumed that the giraffe evolved from an ancestral creature (let us call it species 1) which experienced a chance genetic mutation making its neck grow abnormally long. The long-necked offspring (species 2) was able to browse higher up the tree, free from competition, and hence lived to breed offspring the majority of which also had the longer neck and the same competitive advantage. After an unknown but large number of generations, the short-necked species 1 type creatures had all died out because, among other reasons, they were subject to much more intense competition from the species 2 type creatures.

At some later time, another species 2 type creature experienced another lucky mutation which made its legs grow abnormally long, thus increasing the giraffe's already substantial height advantage and enabling it to browse even higher up the tree. This mutation was again better fitted to its mode of life and hence it flourished with many offspring, the majority also having long legs (let us say species 3).

In time, species 3 became so successful that species 2 became extinct. Any further beneficial mutations took place in a gene-pool that now carried predominantly long-necked and long-legged genes. From time to time a chance mutation for, say, short-leggedness might crop up, but this would not assist the giraffe to breed more successfully (in fact it would seems to be counter-productive), so it would not be selected by nature. However, further beneficial mutations causing further lengthening of the neck or legs would increase the giraffe's ecological edge over other browsers.

A similar sort of scenario can be written for any creature with a distinctive anatomy or way of life. One can picture the shark, perfectly streamlined with its razor-like teeth, evolving by similar processes from some more general fish-like creature. The result of this process over historical time is that practically all creatures today are specialists or occupy vertical ecological niches.

That in general outline is the Darwinian evolutionary scenario. Now look a little more closely at some of the assumptions it contains. The species 2 giraffe fares better than its ancestor, species 1, and becomes the dominant type because its height gives it a feeding advantage. Really? What advantage? Unless it was possible for species 1 to get a living browsing trees at its existing height it clearly would not exist at all – whether you accept the Darwinian view or any other. Its ability to get this living is not affected in any way by the appearance of a taller creature which can exploit food previously out of reach. Perhaps then the advantage species 2 gains is that its new height enables it to exploit virgin territory no other browser can reach? But hold on. Large tracts of Africa both today and in the past are covered with millions of acacia trees, many of which are not eaten by creatures of any sort. There is no need to go up in the air to get a square meal – merely to move a few feet to right or left.

If food were rationed or could be gained only by an animal

with a built-in advantage, then the giraffe would indeed benefit from its fortuitous mutation in the Darwinian sense. But foliage is not in fact usually rationed in that kind of way. (As a matter of fact, the way in which giraffes crop acacia leaves tends to promote subsequent growth, rather than diminish the available supply.) The same is true in the sea where many more potential prey die of old age than are caught and killed by the shark. The so-called ecological advantage is an illusion.

Moreover, once the giraffe has moved higher up the tree, how is it in competition with its short-necked grandparents who are browsing as usual at the lower level? Indeed the higher the giraffe goes, the further away it travels from the original hunting ground and the *less* competition there is for its ancestral stock. The logic of the so-called increased competition is simply faulty.

In the cases discussed so far, we have followed an assumption that Darwinists often make: that the lucky mutations that occur and the characteristics they change are connected to matters affecting the animal's survival. But, of course, the overwhelming majority of mutations would be minor matters not concerned with imminent life or death, but with slight improvements. Mutation as it is conceived of by neo-Darwinists is a matter of degree rather than a matter of radical change.

Consider, for example, the question of camouflage in insects. In some species, this characteristic is very highly developed. The insect called the parent bug feeds mainly on birch leaves and is camouflaged to be about the same colour. Its relative the pied shield bug feeds mainly on white dead nettle and has a lightish patch to resemble the nettle's flower. Another relative, the green shield bug, feeds often on hazel trees and beans, and is appropriately coloured and disguised. All these insects show a remarkable adaptation to their way of life. But none of them is *perfect*. They are well camouflaged, but not infallibly so. If you hunt closely for them, you will find them. And it is a fair bet that their predators will in fact be hunting for them. The significance of this fact is that their protection is a matter of degree. Moreover, it also varies depending on exactly where they are when a predator appears – they may like hazel leaves or whatever, but they cannot spend their entire lives on a hazel leaf. This being the case, they do not possess a decisive and overwhelming advantage over the creature from which they evolved – so why did they evolve? And what happened to the ancestor which was, like them, well adapted but not perfectly adapted? Where are the cheetahs that run at only 50 miles an hour instead of 60? Where are the giraffes only 12 feet tall instead of 17 feet?

Darwinists have recently taken to explaining an important part of the supposed natural selection process in terms of an 'arms race' between competitors: as the prey becomes faster and keener-eyed, so the predator is compelled (by 'selection pressure', of course) to become faster and keener-eyed too – or else lose its livelihood. For instance, the cheetah can run at 60 miles an hour over short distances and is the fastest land animal. It preys mainly on members of the antelope family, of which some members, such as Thomson's gazelle, can run at 50 miles an hour. This seems very persuasive until you apply some analytical thinking to the matter and examine the data from the natural world.

The antelope is fleet of foot – indeed its name is synonymous with speed – but there are many members of the same family (the Artiodactyla) populating the African plains which are just as tasty and much slower. The cheetah is not compelled to eat only the high-speed Thomson's gazelle any more than we are compelled to eat spaghetti bolognaise every night. The cheetah also has plentiful game in the form of the cape buffalo (maximum speed 35 miles an hour), the giraffe (30 miles an hour), the warthog (30 miles an hour) and even the odd sheep or camel (10 miles an hour or so). In fact the cheetah is faster than every other animal, whatever its speed, and so can eat anything it likes.

What natural compulsion is there for the cheetah to engage in an evolutionary arms race? And why should the hypothetical slower animal from which the 60–mile-an-hour cheetah evolved not continue in existence dining as always off the slower members of the antelope family?

The trouble with Darwinism is that it is a little too perfect. Whenever a creature experiences a lucky mutation which aids its way of life, it seems to carry on getting just that kind of mutation. Leaf insects or stick insects look very like leaves and sticks. They must have had successive strokes of luck all driving them in the same genetic direction and no other.

According to Darwinists it is the giraffe and only the giraffe – already abnormally long-necked – that experiences further long-neck mutations, but not other creatures which might benefit equally well. A long neck would benefit any tree-browsing animal and hence should be selected by nature. One would expect to see many other giant animals – and indeed giant men. But only the giraffe has this distinctive trait.

It is asking too much to insist that the giraffe – and only the giraffe – would continue to experience the same kind of mutation, unless the giraffe is uniquely genetically predisposed to experience such mutations. And if this is indeed the case, then it is not chance that has given the giraffe its long neck but some other genetic mechanism of which we presently know nothing.

CHAPTER 20

The Fish that Walked

G. K. Chesterton tells us that:

> God made the wicked grocer
> For a mystery and a sign
> That men might shun the awful shops
> And go to inns to dine.

If God exists and if, as Chesterton thought, he exercised a sense of humour when creating the world, then he must surely have created that mysterious and extraordinary creature the coelocanth to provide mankind with a little light entertainment on wet Sunday afternoons. The story of the coelocanth is worth recounting if only because it reminds us how easy it is for science to get things wrong.

Like most human affairs, science is prone to extraordinary coincidences. On Saint Valentine's Day in 1876 for example, two men walked into the United States Patent Office, each with the same invention under his arm. Alexander Graham Bell and Elisha Gray both filed patents for the telephone that February day, giving rise to a protracted law suit over who had priority – an honour that the courts, and the history books, have awarded to Bell.

Hardly had the dust settled on that law suit when, a decade later in 1886, Charles Hall in the United States and Paul Héroult in France simultaneously but independently devised the electrolytic method for producing aluminium on a commercial scale. It is easy to dismiss these coincidences on the grounds that people working in similar technical fields are likely to come up with similar results. That the coincidences are, so to speak, 'rational coincidences'.

There is, though, another kind of coincidence – the kind of wholly irrational and unpredictable event which Carl Jung termed 'synchronistic' and which has an almost mystical quality. Two of the inventors mentioned earlier, Hall of America and Héroult of France, as well as making the same discovery in the same year of 1886, were both born in 1863 and both died in 1914 – a coincidence which reason is powerless to explain.

For those who, like me, are collectors of coincidences, the Darwinian theory of evolution is a veritable goldmine of improbable events, but few of these incidents have proved quite so embarrassing as a discovery by fishermen off the coast of Africa in 1938 which resulted in the resurrection of a long-dead witness for the prosecution against Darwin – the ghost of the fish that walked.

To appreciate the full significance of the fishermen's strange haul, it is necessary first to go back almost exactly a century earlier, to the survey vessel HMS *Beagle* and to the young Charles Darwin returning home from his epic three-year voyage of natural history discovery in 1836.

On board the *Beagle*, surrounded by fossil remains from distant continents, Darwin began to contemplate the idea of evolution from simple organisms to more complex ones, under the hidden hand of natural selection. The difficulty this has led Darwinists into, as we have seen, is the failure to find any transitional species in the fossil record – or as the newspapers were later to dub them, the 'missing links' in the chain of life.

The missing links looked for were not merely human, but included every part of the animal kingdom: from whelks to whales and from bacteria to bactrian camels. Darwin and his successors envisaged a process beginning with simple marine organisms living in ancient seas, progressing through fishes, to amphibians – living partly in the sea and partly on land – and hence on to reptiles, mammals and eventually the primates, including man. But although each of these classes is well represented in the fossil record, no one has yet discovered a fossil creature that is indisputably transitional between one species and another species. Not a single undoubted 'missing link' has been found in all the exposed rocks of the Earth's crust despite the most careful and extensive searches.

This is a difficulty because, if life has evolved in the way that Darwin proposed, there should be many millions of transitional species – invertebrates with rudimentary backbones; fish with incipient legs; reptiles with half-formed wings; and so on. Indeed, given a theory which postulates continuous random genetic mutation, and hence a continuous spectrum of life forms, constantly evolving to become better and better adapted, such specimens should be the rule rather than the exception. Life itself should be boldly innovative, rather than cautiously conservative.

At first the lack of missing links could be put down to the fact that much of the world remained unexplored and Darwin himself expressed the hope that further exploration would turn up the missing fossils. But the hope gradually faded until it became clear that palaeontology had accumulated an almost unmanageably rich collection of specimens but that the fossil record nevertheless continued to be made up mainly of gaps.

By the time the First World War had ended and the new century was under way it had become abundantly clear that earlier hopes of finding fossils to fill in the many gaps were wearing rather thin, and that further exploration and collecting were merely adding more of the same sort of fossils that were already known and catalogued. Museum departments therefore turned their attention to making sense – that is to say, evolutionary sense – out of the millions of specimens they already had in their store rooms.

These desk researchers naturally looked to comparative anatomy as their guide and focused much of their attention on the major question of the transition from the era of exclusively marine life to that of life on the land; for they instinctively foresaw that if they could provide satisfactorily detailed evidence of this – the first and most important of all transitions – they would provide decisive evidence in favour of the neo-Darwinist model.

Much debate ensued in the palaeontology departments of the world's natural history museums as anatomists examined and rejected, one after another, candidates for the progenitor of all terrestrial life: the fish that had, after millions of years of life in the sea, finally crawled and flapped gasping onto the mud of some ancient estuary to lay its eggs.

The material they had from which to choose was vast. Of all fossils, those of marine creatures are by far the most plentiful because of their greater populations compared with terrestrial

animals and because of more favourable conditions of preservation in ocean sediments. But certain fundamental requirements
were logically obvious from the start. The candidate would be
found amongst the 'bony' fishes rather than amongst those with
merely a flexible cartilaginous skeleton. It must have a well-
developed bony skull. And, most important of all, it must have
four fleshy fins, supported on bony growths, to assist its adven-
tures on land, and from which the four-limbed pattern of life
could have evolved.

These requirements narrowed the field considerably and with
a previously unheard of unanimity, the experts agreed that they
had found their fish. At last the cases of the various claimants
had been examined and the impostors rejected; the pedigree
and credentials of the successful candidate were prepared, and
he was smartened up for presentation to his waiting public. The
'press was squared, the middle classes all prepared', as Hilaire
Belloc observed of a young hopeful in somewhat similar
circumstances.*

The fish that had walked, it was confidently announced, was
of the Crossopterygian (or bony-skulled) class, and more
specifically was a Rhipidistian (or lung-fish). Regrettably, the
fish in question was long-extinct along with all its close relatives, but its anatomy was well known from hundreds of specimens found throughout the fossil record in many parts of the
world, right up to its extinction at about the same time as the
dinosaurs died out, in the Cretaceous period of the Earth's
history.

One particular example of the ancestral fish gave palaeontologists abundant fossil material to study – a fish of the genus
Coelocanthus. Coelocanths had been found in places as far apart
as New Jersey, Greenland, Bavaria, Spitzbergen, Brazil and at
several places in Britain. The coelocanth had been described by
the pioneer palaeontologist Gideon Mantell in the early nineteenth century and had been illustrated by Darwin's champion,
Thomas Huxley, in 1866.

Specimens of the fish had been preserved in ancient rocks in
fine detail and its anatomy had been well studied and catalogued. And it was its anatomical features – plus a little intelligent

---

* Lord Lundy, who was 'destined to be, the next Prime Minister but three.'

guesswork – that prompted such rare unanimity among its authors. The fish and its relatives had flourished during the Devonian period some 350 million years ago, before declining to a dignified end. But before expiring, it had managed to flap onto the estuarine mudflats with the aid of its embryonic limbs, and give birth to a hopeful new generation of creatures able to exploit the land – truly a Columbus among marine organisms and a worthy progenitor of the human race.

The announcement of the discovery of the 'missing link' was one of Fleet Street's earliest scientific scoops and though the readers of the popular dailies couldn't tell a coelocanth from a breakfast kipper, the public imagination was fired by the discovery. The British Museum of Natural History mounted a display and parties of schoolchildren pressed their noses against the hallowed glass cabinets of South Kensington in pursuit of merit marks from approving schoolteachers.

Presumably those responsible for filling the glass cabinets, and the minds behind the noses pressed against them, permitted themselves a moment of self-congratulation. If so, it was short-lived. For at precisely this moment, the most astonishing and irrational coincidence intervened.

Fishermen trawling the waters off East London on the coast of Africa in 1938 found a strange-looking fish in their nets. The decomposing – and by now highly aromatic – remains of the fish were examined by the curator of the East London Museum, Margaret Courtenay-Latimer, and by Professor J. C. B. Smith of Rhodes University, South Africa, who identified it as a living specimen of the coelocanth.

The strange catch was a 'living fossil' and the aroma of stinking fish which attended its discovery must have been poetically inspired by the goddess of coincidence (or perhaps Nemesis). It was no less poetic to those scientists who recalled the indecent haste with which the safely extinct creature had been dragged unceremoniously from the museum store-room to be awarded the posthumous title of Father of Terrestrial Life.

It soon became evident from examining the real article – rather than the imaginatively hopeful museum reconstruction – that the coelocanth was a poor choice for the 'missing link' between marine and terrestrial life. Its four fins are much like those of any other fish and are no more suitable for supporting its weight on land, or of giving rise to amphibious limbs, than those of a fairground goldfish. There is, too, the awkward fact that the coelocanth lives at such great depths in the ocean (up to 200 metres) that it explodes due to decompression when brought up to the surface – a slightly ticklish handicap for a coloniser of the land. In 1986 the coelocanth was observed by underwater TV cameras in its natural habitat by Hans Fricke of the Max Planck Institute for Animal Behaviour. Unsurprisingly, the coelocanth does not stroll on the sea-bed with its fins as supposed, but swims through the water just like any other fish.

Despite its limitations being obvious to anyone who sees a coelocanth (a specimen is preserved and on display in London's Natural History Museum), the publicity surrounding the discovery of a 'living fossil' – far from acting to discredit Darwinism – merely served to reinforce the prestige of science in the public's mind. The only aspect of the affair to stick in the public memory was that evolutionists must be wonderfully clever to even know about a fish before one had been actually caught!

Meanwhile back in the bone departments, the innocent coelocanth was stripped of its titles and dignities in a purely private ceremony. The official line today is that the coelocanth was merely an evolutionary dead-end and some other creature – possibly *Eusthenopteron* – holds the coveted 'missing link' title. *Eusthenopteron*, too, is supposed to be extinct – let us keep our fingers crossed and hope that this time, it stays dead.

This chapter – no more than a piece of fun – might be subtitled 'a cautionary tale' because its story holds a number of lessons both for those who elect to believe in the synthetic or neo-Darwinist theory of evolution and, equally, for those who do not believe in it.

The tale of the 'fish that walked' is a cautionary tale in more ways than one. It cautions us against blind acceptance of the intellectual appeal of an elegant theory, and equally against uncritical acceptance of the intellectual authority of those whom we as a community pay to do our difficult thinking. Scientists are today's Magi or wise men. One of their main functions is to satisfy public curiosity about natural events, but being only human, scientists are sometimes driven to their conclusions by the weight of public demands for knowledge, rather than led to them by the weight of evidence.

When, for instance, J. J. Thomson discovered the electron in 1897, great public and academic interest was aroused and

Thomson was thereafter besieged with demands from students and members of the public wanting to know, 'What is an atom like?' Under such pressure, Thomson hazarded the speculation that an atom resembles an apple with the electrons embedded inside the nucleus like pips – an idea we now know to be false. No one would blame the scientist for entertaining a hypothesis that later proves to be false: indeed that is how science proceeds. But in the case of atomic science, the subject matter is continually present before us for further investigation. The tracks of atomic particles are visible to all in the cloud chamber and errors of theory may be corrected by further observation.

Questions concerning the origin of life, though, are a different matter, for past biological events are no longer available for observation and, regrettably, the tracks they have left are obscure. The traces of biological history that do remain present a vast and often puzzling picture, a picture that can be grasped only through the construction of suitable models. The neo-Darwinist theory is perhaps the most elegant and powerful model ever constructed in the life sciences. But like all models of the real world, it has ultimately reached a point where it is no longer able to contain the data it seeks to explain.

Postscript

One of my earliest discoveries in researching this book was that evolution theory has an almost primeval power to promote heated controversy wherever the subject is raised, even amongst the scientific community who might be expected to be a cool-headed and rational group of people.

Whenever I have discussed with friends and colleagues the ideas contained in this book, their first reaction has invariably been to ask some pointed questions like: 'What is *your* theory of evolution?' and 'What do you *really* believe?' Many of these questions are at root subjective matters whose answers have no place in a book that is an attempt to report objectively on scientific discoveries and controversies.

However, I fully accept that because of the highly contentious nature of this subject matter many readers may well want to know the answers to these and other personal questions in order to know exactly 'where I am coming from' and thus to evaluate whatever bias may be present in my reporting.

The following are my honest answers to those questions that have been put to me by the informed people with whom I have discussed 'the facts of life'.

Q. *Are you a covert creationist, advocating a short time-scale for the Earth's history and casting doubt on Darwinism because of your religious beliefs?*

A. I am not a creationist and do not hold any religious convictions. I can find no scientific or logical reason to believe or disbelieve in a creator and I remain open-minded on the question. Equally, at present I know of no entirely convincing scientific evidence for a natural explanation of the origin of Earth and life. As I have no evidence one way or the other, I have to say I don't know how the world and living things came into being.

I do not belong to any church or religious group, any

academic institution (or indeed any other such group), but greatly prize my independence of thought and action as a freelance writer.

I think that those who believe in a creator, and those who believe in Darwinism, do so as an act of faith or belief, because they find it intellectually or emotionally repugnant to acknowledge and live with such open-minded ignorance.

Q. *What is your theory of evolution?*

A. If chance is not the engine of evolution – and there is a mountain of evidence indicating it is not – then on the face of it, we are thrown back virtually to medieval times in searching for a new foundation for biology – a frightening thought.

I do not believe things are as bad as they seem. There have been many similar occasions in the history of science when research has reached an apparent impasse, when things have looked black, but the very act of breaking the stalemate has been a creative achievement bringing much valuable new knowledge.

Having considered all the evidence, I am convinced we are on the brink of a major breakthrough linking the discoveries of quantum mechanics to the form and processes of biological structures. The first step in this breakthrough, I believe, will be the recognition by science that certain biological phenomena which have been known for centuries, but have been marginalised out of sight and out of mind by reductionist thinking (exemplified by Darwinism), are quantum phenomena taking place at the macroscopic level.

Q. *What should we teach in schools and universities?*

A. This is the toughest question of all. I believe that Darwinism is a philosophy rather than science and in higher education could be taught as such – perhaps biology as an academic subject should be divided into the experimental and the theoretical. Darwinism could be one of the theories offered.

In primary and secondary education, I believe that we simply have to come clean with students and tell them the truth. That the evidence for overall evolution is fragmentary and often contradictory; that we have no satisfactory model of a mechanism to drive the evolutionary process; that there is some evidence for the inheritance of acquired characteristics and some evidence for a Darwinist mechanism but that neither body of evidence is by itself compelling.

I believe we must also come clean over human evolution and stop filling the classroom with over-imaginative 'restorations' and 'reconstructions' of ancestors that look part-ape and part-human in defiance of the evidence. We must also point out the serious difficulties in dating techniques and admit that we do not have a satisfactory model for the formation of the Earth.

Q. *Are you on some other kind of covert intellectual crusade? Surely Darwinism is so self-evidently true to any rational person who studies the evidence that only some kind of nut can deny it.*

A. It certainly looks that way to most thinking people. The structure of the theory seems quite solid. It is only when you start asking awkward questions and start turning over stones that you begin to find anomalies. And the evidence I cite in this book comes from reputable scientists, usually people in substantial academic positions. Take another look at the evidence on dating in Chapters 3 and 4 – for instance the Hawaiian volcanic lavas dated as *billions* of years old by the potassium-argon method that are actually 190 years old. This is not simply a slip-up by otherwise conscientious scientists. It is the result of a systematic defect in the dating method caused by a hitherto unrecognised technical factor which many Darwinists simply refuse to acknowledge.

Far from being on any crusade for my own ideas or beliefs, my primary aim in writing this book is to sound a cautionary note about the extent to which ideological Darwinism has replaced scientific Darwinism in our educational system. My message is that the world is full of people who want you to believe in their 'ism' –

Darwinism, Marxism, Freudianism, and the rest. Don't accept anything they say unless they can substantiate it with scientific evidence, however persuasive their arguments.

Q.   *Aren't you just trying to stir up trouble? Why don't you leave scientists to get on with their jobs, and get on with your own?*

A.   Almost all scientists today earn their living from the public purse in one way or another. In effect, we the community employ scientists to tackle the difficult task of explaining that which we do not understand. This is no easy job, to be sure, and one in which success may depend on luck every bit as much as skill and judgement. Because it is a difficult job, a tacit understanding has arisen that it would be bad form or unseemly to criticise science or scientists in any serious way, as if they were a bank who got their sums wrong or a grocer who forgot to deliver the sausages.

I reject this tacit consensus. I am a customer for the scientific service that we pay scientists to provide, and I have a customer complaint: I am not satisfied with the answers they have provided on the mechanism of evolution and I want to them to go back to their laboratories and think again.

I am not being mean-minded to biologists. I am simply providing some quality-control feedback. I believe it is high time that consumerism found a voice in the public sector as effectively as it has in industry and commerce. And I do not accept the convention that scientists may be criticised only by their peers.

Finally, I believe that science and reason – tempered by intuition – offer the only real hope of discovering answers to these baffling questions and I wholeheartedly support the Western scientific method of enquiry. I am, though, concerned that many people, including some scientists, pay lip service to this idea while thinking and acting like intellectual Stalinists.

There is a strong streak of intellectual arrogance and intellectual authoritarianism running through the history of Darwinism, from Thomas Huxley and Charles Darwin (both openly racist) through to Sir Julian Huxley, the principal architect of the neo-Darwinist theory in the twentieth century, who publicly advocated that people who were genetically abnormal (such as those mentally and physically handicapped by heredity) should be sterilised in order to relieve society from having to care for their offspring.

This authoritarian streak is still present in some Darwinists today and is denoted by the outrage and indignation with which they greet any reasoned attempt to expose the theory to debate and to the light of real evidence. I believe this reaction is caused by the psychological phenomena described by Thomas Kuhn and Leon Festinger, referred to in Chapter 14, rather than because of any malicious intention on their part. But I also believe that the effect is the same as if it was intentionally malicious and hence should be resisted by all people who prize their individuality and independence of mind.

Darwinism has never had much appeal for science outside of the English-speaking world, and has never appealed much to the American public (despite being enthusiastically taken up by the US scientific establishment in the past). However, its ascendancy in science, even in Britain, has been waning for several decades as its grip has weakened in successive areas: geology; palaeontology; embryology; comparative anatomy. Now even geneticists are beginning to have doubts. It is only in mainstream microbiology and zoology that Darwinism retains serious enthusiastic supporters.

As scientists begin to drift away from neo-Darwinist ideas in growing numbers, the revision of Darwinism at the public level is long overdue, and is a process that I believe has already started.

References

CHAPTER 1

1. Francis Crick, 1970.
2. Jacques Monod, 1972.
3. F. M. Broadhurst, 1964.
4. S. K. Vsekhsviatsky, 1964, 1966.
5. See Chapter 15.

CHAPTER 2

1. Stanley Miller, 1953.
2. Murray Eden, 1967.
3. Francis Crick, 1981.
4. John Thackray, 1980.
5. E. Barghoorn, 1971.
6. H. D. Pflug and H. Jaeschke-Boyer, 1979.
7. D. Lowe, 1980.
8. Quoted in Taylor, 1983.
9. Van Eysinga, 1975.
10. Arthur Holmes, 1960.
11. Henry Faul, 1959, 1960, 1978.
12. See, for instance, Harold Levin, 1978.
13. John Thackray, 1980.
14. Gavin de Beer, 1970.
15. Harold L. Levin. 1978.
16. W. H. Bradley. 1929.

CHAPTER 3

1. Ronald Millar, 1972
2. Johann Scheuchzer, 1726.
3. For example, D. N. Wadia, 1966 edn, on the Siwalik Hills beds.
4. Charles Lyell, 1833.
5. W. F. Libby, 1955.
6. Richard E. Lingenfelter, 1963.
7. Hans Seuss, 1965.
8. V. R. Switzer, 1967.

9. Melvin Cook, October 1968.
10. Ibid.
11. M. S. Kieth and G. M. Anderson, 1963.
12. R. W. Fairbridge, 1984.

CHAPTER 4

1. John L. Mero, 1965.
2. James Lovelock, 1987 edn.
3. Frederick Jueneman, 1972.
4. Henry Morris, 1985 edn.
5. Melvin Cook, 1966.
6. Melvin Cook, 1957
7. James Lovelock, 1987 edn.
8. Henry Morris, 1985 edn.
9. Melvin Cook, 1966.
10. C. S. Noble and J. J. Naughton, 1968.
11. G. J. Funkhouser and J. J. Naughton, 1968.

CHAPTER 5

1. Hans Petterson, 1960.
2. Henry Morris, 1985 edn.
3. S. K. Vsekhsviatsky, 1964, 1966.
4. Thomas Barnes, 1983 edn.
5. Melvin Cook, 1966.

CHAPTER 6

1. Sir Leonard Woolley, 1950.
2. Alfred Wegener, 1924.
3. P. M. Hurley, 1968.
4. Melvin Cook, 1966.
5. Ibid.
6. Charles Hapgood and James Campbell, 1958.

CHAPTER 7

1. Stephen Jay Gould, 1990.
2. R. W. Gallois, 1965.
3. C. O. Dunbar and John Rogers, 1957.
4. Stuart Nevins, 1971.
5. John L. Mero, 1965.
6. V. I. Sozansky, March 1973.
7. Omer Roup, December 1970.
8. I. Velikovsky, 1950.

CHAPTER 8

1. S. E. Hollingsworth, 1962.
2. Melvin Cook, 1966.
3. F. M. Broadhurst, 1964.

CHAPTER 9

1. I. Velikovsky, 1973 edn.
2. Alfred De Grazia et al. 1966.
3. I. Velikovsky, 1973 edn.
4. Helmut de Terra and T. T. Paterson, 1939.
5. Hugh Miller, 1869 edn.
6. William Buckland, 1836.
7. Harry S. Ladd, 1959.
8. C. O. Dunbar, 1960.
9. D. N. Wadia, 1966 edn.
10. Edwin Colbert, 1968.
11. Edwin Colbert, 1965.

CHAPTER 10

1. George Simpson, 1950.
2. Ibid.
3. Garrett Hardin, 1961.
4. George Simpson, 1950.
5. Sankar Chatterjee, 1991.
6. Charles Darwin, 1901 edn.
7. Charles Darwin, 1902 edn.
8. J. G. O. Smart et al. 1966
9. J. H. Callomon, 1968.
10. Raymond C. Moore (ed.), 1957.
11. Norman Macbeth, 1974.
12. Ernst Mayr, 1953.
13. See also Adrian Desmond, 1975.
14. W. R. Thompson, 1956
15. Keith Simpson, 1978.

CHAPTER 11

1. Charles Darwin, 1902 edn.
2. Julian Huxley, 1963.
3. C. H. Waddington, 1960.
4. Charles Darwin, 1902 edn.
5. G. G. Simpson, 1967.
6. Julian Huxley, 1963.

7.   G. G. Simpson, 1964.
8.   Sir Gavin de Beer, 1970.
9.   Theodosius Dobzhansky, 1984.

CHAPTER 12

1.   Ernst Mayr, 1963.
2.   Ernst Mayr, 1970.

CHAPTER 13

1.   Julian Huxley, 1963.
2.   Jacques Monod, 1970.
3.   Entry on Human Genetics in *Encyclopaedia Britannica*, 1984.
4.   William Paley, 1828.
5.   Ronald Fisher, 1930.
6.   Richard Dawkins, 1986.
7.   Ibid.
8.   Francis Crick, 1981.
9.   Charles Darwin, 1902 edn.
10.   David Norman, 1985.

CHAPTER 14

1.   Thomas Kuhn, 1962.
2.   Charles Darwin, 1901 edn.
3.   Ernst Weidersheim, 1895.
4.   S. R. Scadding, May 1981.
5.   G. G. Simpson, 1951.
6.   Ernst Haeckel, 1866.
7.   A. J. White, 1989.
8.   Sir Gavin De Beer, 1964.
9.   Ernst Mayr, 1960.
10.   G. G. Simpson, 1967.
11.   George Simpson, Pittendrigh and Tiffany, 1958.
12.   Arthur Koestler, 1967.

CHAPTER 15

1.   Ronald Millar, 1972.
2.   Ernst Mayr, 1963.
3.   B. R. Neufeld, 1975.
4.   Edwin G. Conklin, 1943.
5.   Andrew Scott, 1988.
6.   Francis Crick, 1966.
7.   Willam Shrive, 1982.

CHAPTER 16

1. Ernst Haeckel, 1876 edn.
2. Ernst Haeckel, 1899 edn.
3. A. J. White, 1989.
4. A. S. Romer, 1966.
5. A. J. Kelso, 1974.
6. Richard E. Leakey, 1981.
7. A. J. E. Cave, and W. L. Strauss, 1957.
8. Raymond Dart, 1925.
9. Sir Solly Zuckerman, 1954.
10. Philip Tobias, 1967.
11. A. J. White, 1989.
12. See John Reader, 1981.

CHAPTER 17

1. Jean-Baptiste de Lamarck, 1809.
2. Alan Durrant, 1958, 1962.
3. J. Hill, 1965.
4. C. A. Cullis, 1977.
5. August Weisman, 1893.
6. Francis Crick, 1970.
7. Howard Temin, 1976.
8. Edward Steele, 1979.
9. Reg Gorczinski and Edward Steele, 1977.
10. Barry Hall, September 1990.
11. Richard Goldschmidt, 1940.
12. Stephen Jay Gould, 1977.
13. Svante Arrhenius, 1908.
14. Sir Fred Hoyle and Chandra Wickramasinghe, 1981.
15. Sir Fred Hoyle, 1983.
16. Francis Crick, 1981.

CHAPTER 18

1. T. Humphreys, 1963.
2. A. A. Moscona, 1959.
3. B. I. Balinsky, 1951.
4. D. J. Pritchard, 1990.
5. C. H. Waddington, 1974.
6. *Physical Review*, March 1990.
7. Einstein, Podolsky and Rosen, 1935.
8. David Bohm, 1980.
9. Hans Dreisch Presidential Address to the Society for Psychical
    Research, 1926.

10. Sir Alister Hardy, 1949, 1950, 1953.

CHAPTER 19

1. Garrett Hardin, 1960.
2. Sir Gavin De Beer, 1964.
3. M. and A. Wolsky, 1976.
4. E. Denton, 1971.
5. Richard Goldschmidt, 1940.

Bibliography

ARRHENIUS, Svante, 1908, *Worlds in the Making*, London.

BAKKER, Robert T., 1971, *Dinosaur physiology and the origin of mammals* in *Evolution*, 25, pp. 636–58.

BALINSKI, B. I., 1951, *On the eye cup-lens correlation in some South African amphibians* in *Experimenta*, 7, 180–1.

BARGHOORN, E., 1971, *The oldest fossils* in *Scientific American*, 224(5)30.

BARNES, Thomas G., 1983 edn, *Origin and destiny of the Earth's magnetic field*, Institute for Creation Research, San Diego.

BLACK, M., 1953, *The constitution of chalk*, Report of lecture in *Proceedings of the Geological Society*, No. 1499, lxxxi–vi.

BOHM, David, 1980, *Wholeness and the Implicate Order*, Routledge & Kegan Paul, London.

BRADLEY, W. H., 1929, *The varves and climate of the Green River epoch*, Professional paper 158, Unites States Geological Survey.

BRIGGS, D., and Walters, S. M., 1984 edn, *Plant Variation and Evolution*, Cambridge University Press.

BROADHURST, F. M., 1964, *Some aspects of the palaeoecology of non-marine faunas and rates of sedimentation in the Lancashire coal measures* in *American Journal of Science*, Vol. 262, p. 865.

BUCKLAND, William, 1823, *Reliquiae Diluvianae*, John Murray, London.

BUCKLAND, William, 1836, *Geology and Mineralogy Considered with reference to Natural Theology*, Pickering, London.

CAIRNS, J., Overbaugh, J., and Miller, S., 1988, *The origin of mutants*, in *Nature*, 335: 142–5.

CALLOMON, J. H., 1968, in P. C. Sylvester-Bradley (ed.), *The Geology of the East Midlands*, Leicester University Press.

CAVE, A. J. E. and Strauss, W. L., 1957, *Pathology and posture of Neanderthal man* in *Quarterly Review of Biology*, Vol. 32, pp. 348–63.

CHATTERJEE, Sankar, July 1991, *Protoavis texensis* in *Philosophical Transactions of the Royal Society*.

COLBERT, Edwin, 1965, *The Age of Reptiles*, W. W. Norton, New York.

COLBERT, Edwin, 1968, *Men and Dinosaurs*, E. P. Dutton, New York.

CONKLIN, Edwin G., 1943, *Man Real and Ideal*, Scribners, New York.

COOK, Melvin A., January 1957, *Where is the Earth's radiogenic helium?* in *Nature*, Vol. 179, p. 213.

COOK, Melvin A., 1966, *Prehistory and Earth Models*, Max Parrish, London.

COOK, Melvin A., October 1968, *Do radiological clocks need repair?* in *Creation Research Society Quarterly*, Vol. 5, p. 70.

COX, L. R. (ed.), 1967 edn, *British Mesozoic Fossils*, British Museum of Natural History, London.

CRICK, Francis, 1966, *Of Molecules and Men*, University of Washington Press, Seattle.

CRICK, Francis, 1970, in *Nature*, 227, 561.

CRICK, Francis, 1981, *Life Itself*, Macdonald. London.

CRICK, Francis, and Orgel, Leslie, 1973, *Directed panspermia* in *Icarus*, Vol. 19, p. 341.

CULLIS, C. A., 1977, *Molecular aspects of the environmental induction of heritable changes in Flax*, in *Heredity*, Vol. 38, pp. 129–54.

DART, Raymond, 1925, *Australopithecus africanus: The Man-Ape of South Africa* in *Nature*, 115: 195–9.

DARWIN, Charles, 1901 edn, *The Descent of Man*, John Murray, London

DARWIN, Charles, 1902 edn, *The Origin of Species*, John Murray, London.

DAWKINS, Richard, 1986, *The Blind Watchmaker*, Longman, London.

DE BEER, Sir Gavin, 1964, *Atlas of Evolution*, Nelson, London.

DE BEER, Sir Gavin, 1970, *A Handbook of Evolution*, British Museum of Natural History, London.

DE GRAZIA, Alfred, R. E. Juergens, and L. C. Stecchini, 1966, *The Velik-ovsky Affair*, Sidgwick & Jackson, London.

DE TERRA, Helmut, and Paterson, T. T., 1939, *Studies on the Ice Age in India and Associated Human Cultures*, Harvard University Press.

DENTON, E., 1971, *Reflectors and fishes* in *Scientific American*, 224(1)64.

DESMOND, Adrian, 1975, *The Hot Blooded Dinosaurs*, Blond & Briggs, London.

DOBZHANSKY, Theodosius, 1941, *Genetics and the Origin of Species*, Columbia University Press.

DOBZHANSKY, Theodosius, 1984 edn, *Heredity* entry in *Encyclopaedia Britannica*, 15th edition.

DREISCH, Hans, 1908, *The Science and Philosophy of the Organism*, London.

DREISCH, Hans, 1925, *The Philosophy of Vitalism*, London.

DUNBAR, C. O., 1960 edn, *Historical Geology*, John Wiley, New York.

DUNBAR, C. O., and Rogers, John, 1957, *Principles of Stratigraphy*, John Wiley, New York.

DURRANT, Alan, 1958, *Environmental conditioning of flax* in *Nature*, Vol. 81, pp. 928–9.

DURRANT, Alan, 1962, *The environmental induction of heritable changes in Linum*, in *Heredity*, Vol. 17, pp. 27–61.

EDEN, Murray, 1967, *The inadequacy of neo-Darwinian evolution as a scientific theory*, Massachusetts Institute of Technology conference paper.

EINSTEIN, Albert, Podolsky, Boris, and Rosen, Nathan, 1935, *Can quantum-mechanical description of physical reality be considered complete?* in *Physical Review*, 47, 777.

ELDREDGE, Niles, and Gould, Stephen Jay, 1960, *Phylogenetic Patterns and the Evolutionary Process*, Columbia University Press.

FAIRBRIDGE, R. W., 1964, *African Ice-Age aridity* in A. E. M. Nairn (ed.), *Problems in Palaeoclimatology*, London.

FAIRBRIDGE, R. W., 1984 edn, *Holocene* entry in *Encyclopaedia Britannica*, 15th edition.

FAUL, Henry, 1959, *Doubts of the Palaeozoic time scale* in *Journal of Geophysical Research*, Vol. 64, p. 1102.

FAUL, Henry, 1960, *Geologic time scale* in *Bulletin of the Geological Society of America*, Vol. 71, pp. 637–44.

FAUL, Henry, 1978, *A history of geologic time* in *American Scientist*, Vol. 66, pp. 159–65.

FISHER, Ronald, 1930, *The Genetical Theory of Natural Selection*, Oxford University Press.

FREEDMAN, S., and Clauser, J., 1972, *Experimental test of local hidden variables theories* in *Physical Review Letters*, Vol. 28, p. 938.

FUNKHOUSER, J. G., and Naughton, J. J., July 1968, *Journal of Geophysical Research*, Vol. 73, p. 4606.

GALLOIS, R. W., 1965, *The Wealden District*, Institute of Geological Sciences, London.

GOLDSCHMIDT, Richard, 1940, *The Material Basis of Evolution*, Yale University Press.

GORCZYNSKI, Reg, and Steele, Edward, 1977, *Inheritance of acquired immunological tolerance to foreign histocompatibility antigens* in *Proceedings of the National Academy of Sciences of the USA*, p. 2871.

GOULD, Stephen Jay, 1977, *Ontogeny and Phylogeny*, Harvard University Press.

GOULD, Stephen Jay, 1990, *Wonderful Life: the Burgess Shales and the Nature of History*, Hutchinson Radius, London.

HAECKEL, Ernst, 1876 edn, *The General Morphology of Organisms*, London.

HAECKEL, Ernst, 1876 edn, *The History of Creation*, London.

HAECKEL, Ernst, 1899, *The Last Link*, London.

HALL, Barry G., Sept. 1990, *Spontaneous point mutations that occur more often when advantageous than when neutral* in *Genetics*, Vol. 126, pp. 5–16.

HAPGOOD, Charles, and Campbell, James, 1958, *Earth's Shifting Crust*, Museum Press, London.

HARDIN, Garrett, 1961, *Nature and Man's Fate*, Mentor, New York.

HARDY, Sir Alister, May/June 1950, article in *Journal of the Society for Psychical Research*, London.

HARDY, Sir Alister, 1953, article in *Proceedings of the Society for Psychical Research*, Vol. 50. London.

HARDY, Sir Alister, 1965, *The Living Stream*, Collins, London

HILL, J., 1965, *Environmental induction of heritable changes in Nicotiana rustica* in *Nature*, Vol. 207, pp. 732–4.

HOLLINGSWORTH, S. E., 1962, *The climatic factor in the geological record* in *Quarterly Journal, Geological Society of London*, Vol. 118, p. 13.

HOLMES, Arthur, 1947, *The construction of a geological time scale* in *Transactions of the Geological Society of Glasgow*, Vol. 21, pp. 117–52.

HOLMES, Arthur, 1960, *A revised geological time scale* in *Edinburgh Geological Society Transactions*, Vol. 17, Part 3, pp. 183–216.

HOLMES, Arthur, 1965, *Principles of Physical Geology*, Ronald Press, New York.

HOYLE, Sir Fred, 1983, *The Intelligent Universe*, Michael Joseph, London.

HOYLE, Sir Fred, and Wickramasinghe C., 1978, *Lifecloud*, J. M. Dent, London.

HUMPHREYS, T., 1963, *Chemical dissolution and in vitro reconstruction of sponge cell adhesions* in *Developmental Biology*, Vol. 8, pp. 27–47.

HURLEY, P. M., 1968, *The confirmation of continental drift* in *Scientific American*, 218(4)52.

HUTTON, James, 1899 edn, *Theory of the Earth with Proofs and Illustrations*, Archibald Geikie (ed.), Geological Society, London.

HUXLEY, Julian S., 1963 edn, *Evolution – the Modern Synthesis*, Allen & Unwin, London.

HUXLEY, Julian S., A. C. Hardy, and E. B. Ford (eds), 1954, *Evolution as a Process*, Allen & Unwin, London.

JUENEMAN, Frederick, September 1972, *Scientific speculation* in *Industrial Research*, p. 15.

KELSO, A. J., 1974, *Physical Anthropology*, J. P. Lipincott, New York.

KIETH, M. S., and Anderson, G. M., 16 August 1963, *Radiocarbon dating: fictitious results with mollusk shells* in *Science*, p. 634.

KOESTLER, Arthur, 1967, *The Ghost in the Machine*, Hutchinson, London.

KOESTLER, Arthur, 1978, *The Case of the Midwife Toad*, Hutchinson, London.

KUHN, Thomas, 1962, *The Structure of Scientific Revolutions*, Chicago University Press.

LADD, Harry S., 1959, *Ecology, Paleontology and Stratigraphy* in *Science*, Vol. 129, p. 72.

LAMARCK, Jean-Baptiste de, 1809, *Philosophie Zoologique*, Paris.

LEAKEY, Richard E., 1981, *The Making of Mankind*, Michael Joseph, London.

LEVIN, Harold L., 1978, *The Earth Through Time*, W. B. Saunders, Philadelphia.

LEWIS, John (ed), 1974, *Beyond Chance and Necessity*, Garnstone Press, London.

LIBBY, Willard F., 1955 edn, *Radiocarbon Dating*, Chicago University Press.

LINGENFELTER, Richard E., February 1963, *Production of C-14 by cosmic ray neutrons* in *Review of Geophysics*, Vol. 1, p. 51.

LOVELOCK, James E., 1987 edn, *Gaia: A New Look at Life on Earth*, Oxford University Press.

LOWE, D., 1980, *Stromatolites 3400 Myr-old from the Archean of Western Australia* in *Nature*, 284:441.

LYELL, Charles, 1833 edn, *Principles of Geology*, John Murray, London.

MACBETH, Norman, 1974 edn, *Darwin Retried*, Garnstone Press, London.

MANTELL, Gideon, 1822, *The fossils of the South Downs, or illustrations of the geology of Sussex*, Relfe, London.

MANTELL, Gideon, 1825, *On the teeth of the Iguanodon* in *Philosophical Transactions of the Royal Society*, Vol. 115, pp. 179–86.

MAYR, Ernst, 1953, *Methods and Principles of Systematic Zoology*, McGraw Hill, New York.

MAYR, Ernst, 1960, *The emergence of evolutionary novelties* in *Tax*, Vol. 1, pp. 349–80.

MAYR, Ernst, 1963, *Animal Species and Evolution*, Harvard University Press.

MAYR, Ernst, 1964 edn, *Systematics and the Origin of Species*, Dover, New York.

MAYR, Ernst, 1970, *Populations, Species and Evolution*, Harvard University Press.

MERO, John. L., 1965, *The Mineral Resources of the Sea*, Elsevier, London.

MILLAR, Ronald, 1972, *The Piltdown Men*, Gollancz, London.

MILLER, Hugh, 1869 edn, *The Old Red Sandstone*, W. P. Nimmo, Edinburgh.

MILLER, Stanley. L., 1953, *A production of amino acids under possible primitive Earth conditions* in *Science*, Vol. 117, pp. 528–9.

MONOD, Jacques, 1972 edn, *Chance and Necessity*, William Collins, Glasgow.

MOORE, Raymond C. (ed), 1957, *Treatise on Invertebrate Paleontology. Part L – Mollusca*, Geological Society of America and Kansas University Press.

MOOREHEAD, Alan, 1969, *Darwin and the Beagle*, Hamish Hamilton, London.

MORRIS, Henry M. (ed.), 1985 edn, *Scientific Creationism*, Institute of Creation Research, San Diego.

MOSCONA, A. A., 1959, *Tissues from dissociated cells* in *Scientific American*, 200(5)132.

NEUFELD, B. R., 1975, *Dinosaur tracks and giant men* in *Origins*, Vol. 2, pp. 64–76.

NEVINS, Stuart E., 1971, *Stratigraphic evidence of the flood* in *Symposium on Creation III*, Baker Book House, Grand Rapids.

NOBLE, C. S., and Naughton, J. J., October 1968, *Deep ocean basalts: inert gas content and uncertainties in age dating* in *Science*, Vol. 162, p. 265.

NORMAN, David, 1985, *Encyclopaedia of Dinosaurs*, Salamander Books, London.

OPDYKE, N. D., 1968, *Palaeomagnetism of Oceanic Cores* in R. A. Phinney's *History of the Earth's Crust*.

OPDYKE, N. D., and Runcorn, S. K., 1956, *New evidence for the reversal of the geomagnetic field near the Plio-Pleistocene boundary* in *Science*, Vol. 123, No. 3208.

PALEY, William, 1828, *Natural Theology*, London.

PETTERSSON, Hans, 1948–49, *The Swedish deep sea expedition 1947–1948* in *Geological Journal*, Vol. 114, Nos. 4–6 (1948), No. 406 (1949).

PETTERSSON, Hans, February 1960, *Cosmic spherules and meteoritic dust* in *Scientific American*, Vol. 202, p. 132.

PFLUG, H. D., and Jaeschke-Boyer, H., 1979, *Combined structural and chemical analysis of 3800 Myr-old microfossils* in *Nature*, 280, p. 483.

PHINNEY, R. A., 1968, *History of the Earth's Crust*, Princeton University Press.

PRITCHARD, D. J., 1990, *The missing chapter in evolution theory* in *Biology*, 37 (5) 149–52.

READER, John, 1981, *Missing Links*, Collins, London.

ROMER, A. S., 1966 edn, *Vertebrate Palaeontology*, University of Chicago Press, Chicago and London.

ROMER, A. S., 1968, *The Process of Life*, Weidenfeld & Nicolson, London.

ROUP, Omer B., December 1970, *Brine mixing: an additional mechanism for formation of basin evaporites* in *Bulletin, American Association of Petroleum Geologists*, Vol. 54, p. 2258.

SCADDING, S. R., May 1981, *Do vestigial organs provide evidence for evolution?* in *Evolutionary Theory*, Vol. 5, pp. 173–6.

SCHEUCHZER, Johann, 1726, *Homo Diluvii Testis*, Burkli, Zurich.

SCHRODINGER, Erwin, 1955 edn, *What is Life?*, Cambridge University Press.

SCOTT, Andrew, 1988, *Vital Principles*, Blackwell, Oxford.

SHRIVE, William, 1982, entry on *Enzymes* in *Encyclopedia of Science and Technology*, McGraw Hill, New York.

SIMPSON, George G., 1944, *Tempo and Mode in Evolution*, Columbia University Press.

SIMPSON, George G., 1951, *Horses*, Oxford University Press.

SIMPSON, George G., 1964, *This View of Life*, Harcourt Brace & World, New York.

SIMPSON, George G., 1967 edn, *The Meaning of Evolution*, Yale University Press.

SIMPSON, George G., 1969, *Biology and Man*, Harcourt Brace.

SIMPSON, George G., Pittendrigh, C. S., and Tiffany, L. H., 1958, *Life: an Introduction to Biology*, Routledge & Kegan Paul, London.

SIMPSON, Keith, 1978, *Forty Years of Murder*, Harrap & Co., London.

SMART, J. G. O., et al., 1966, *Geology of the country around Canterbury and Folkestone*, HMSO for Geological Survey of Great Britain.

SMITH, Andrew B., 1987, *Fossils of the Chalk*, The Palaeontological Association, London.

SOZANSKY, V. I., March 1973, *Origin of salt deposits in deep-water basins of Atlantic ocean* in Bulletin, American Association of Petroleum Geologists, Vol. 57, p. 590.

SPATH, L. F., 1923–43, *Ammonoidea of the Gault*, parts 1–16, The Palaeontographical Society, London.

STEELE, Edward (ed.), 1979, *Genetic Selection and Adaptive Evolution*, Williams & Wallace, Toronto.

SUESS, Hans E., December 1965, *Secular variations in the Cosmic-ray produced Carbon-14 in the atmosphere and their interpretations* in Journal of Geophysical Research, Vol. 70, p. 5947.

SWITZER, V. R., August 1967, *Radioactive dating and low level counting* in Science, Vol. 157, p. 726.

TAYLOR, Gordon Rattray, 1983, *The Great Evolution Mystery*, Secker & Warburg, London.

TEMIN, Howard M., 1976, *The DNA provines hypothesis* in Science, Vol. 192, p. 1075.

THACKRAY, John, 1980, *The Age of the Earth*, Institute of Geological Sciences, London.

THOMPSON, W. R., 1956, Introduction to *The Origin of Species*, Everyman Library, Dutton, London.

THOMSON, William (Lord Kelvin), 1862, *On the secular cooling of the earth* in Royal Society of Edinburgh Transactions, Vol. 23, No. 1, pp. 157–69.

TOBIAS, P. V., 1967, *Olduvai Gorge 2: The cranium and maxillary dentition of Australopithecus (Zinjanthropus) boisei*, Cambridge.

VAN EYSINGA, F. W. B., 1975 edn, *Geological Time Table*, Elsevier Scientific, Amsterdam.

VELIKOVSKY, Immanuel, 1973 edn (1950), *Worlds in Collision*, Sphere, London.

VELIKOVSKY, Immanuel, 1973 edn (1955), *Earth in Upheaval*, Sphere, London.

VINE, F. J., 1968, *Magnetic Anomalies Associated with Mid-Ocean Ridges* in R. A. Phinney's *History of the Earth's Crust*.

VSEKHSVIATSKY, S. K., 1964, *Physical Characteristics of Comets*, London.

VSEKHSVIATSKY, S. K., 1966, *The Nature and Origin of Comets*, London.

WADDINGTON, C. H., 1960, *Evolutionary adaptation* in *Tax*, Vol, 1, pp. 381–402.

WADDINGTON, C. H., 1974, *How much is evolution affected by chance and necessity?*, in *Beyond Chance and Necessity*, John Lewis (ed), Garnstone Press, London.

WADIA, D. N., 1966 edn, *Geology of India*, Macmillan, London.

WEGENER, Alfred, 1924, *The Origin of Continents and Oceans*, Methuen, London.

WEIDERSHEIM, Ernst, 1895, *The Structure of Man*.

WEISMAN, August, 1893, *The Germ Plasm: A Theory of Heredity*, W. Scott, London.

WHITE, A. J., 1989, *Wonderfully Made*, Evangelical Press, Darlington.

WOLKSY, M. and A., 1976, *The Mechanism of Evolution: a New Look at Old Ideas*, Karger, Basle.

WOOLLEY, Sir Leonard, 1950 edn, *Digging up the Past*, Penguin Books, London.

ZUCKERMAN, Sir Solly, 1954, *Correlation of changes in the evolution of the higher primates*, in Huxley, Hardy and Ford, *Evolution as a Process*.

ZUKAV, Gary, 1979, *The Dancing Wu Li Masters*, Hutchinson, London.

Subject Index

Index of Names

Index of Species